ESTRANGEMENT

Estrangement

America and the World

A CARNEGIE ENDOWMENT BOOK

Edited by

SANFORD J. UNGAR

New York Oxford
Oxford University Press
1985

OXFORD UNIVERSITY PRESS

Oxford New York Toronto
Delhi Bombay Calcutta Madras Karachi
Petaling Jaya Singapore Hong Kong Tokyo
Nairobi Dar es Salaam Cape Town
Melbourne Auckland

and associated companies in
Beirut Berlin Ibadan Nicosia

Published by Oxford University Press, Inc.,
200 Madison Avenue, New York, New York 10016

Library of Congress Cataloging-in-Publication Data
Main entry under title:
Estrangement: America and the world.
"A Carnegie Endowment book."
Bibliography: p. Includes index.
1. United States—Foreign relations—1945– —Addresses, essays, lectures.
I. Ungar, Sanford J.
E840.E87 1985 327.73 85-18733
ISBN 0-19-503707-3

Printing (last digit): 9 8 7 6 5 4 3 2 1
Printed in the United States of America

Acknowledgments

The idea for this book was nurtured from its earliest stages by Tom Hughes and Larry Fabian; it is characteristic of their stewardship of the Carnegie Endowment to sponsor intellectual inquiry and independent research, notwithstanding the risks that may emerge along the way. The idea was taken up, most notably, by my eleven willing collaborators, who gave a great deal of time and substantial energy to their assignments, and by Susan Rabiner and Rachel Toor at Oxford University Press; their prodding and provoking ultimately made a great difference.

Help was provided at various stages by Michael O'Hare, Rosemary Gwynn, Gladys Bostick, Amanda Cadle, Mary Child, Dieuxuan McGeever, Kendra Ensor, Abigail Russell, Chuck O'Boyle, Leslie Weinfield, Vicki Levy, Brad Woodhouse, Jonathan Simon, Ann-Eve Pederson, and Mark Colodny. The library at the Carnegie Endowment, led by Jane Lowenthal, responded to countless desperate pleas, and Sara Goodgame, who is largely responsible for the chronology of postwar events that ends the book, also got us through one deadline after another with great aplomb.

My closest partner in this endeavor was David Weiner, a clear thinker, a persistent skeptic, a superb editor, and a good friend. He deserves a large share of any credit that is due.

Washington S.J.U.
September 1985

Contents

Foreword

THOMAS L. HUGHES, PRESIDENT
Carnegie Endowment for International Peace

> We do not believe in permanent estrangement. . . . Above all, the United States has yet to resolve the contradictions between the traditions of the founding fathers and of Lincoln and the external image it gives of superpower politics.
>
> Indira Gandhi, addressing Americans, 1972

> The relationship between the United States and India has become estranged. That estrangement has [arrived] at a point where India has on many subjects closer relations with the Soviet Union than with the U.S. You might ask the government of India why they prefer to have closer relations with that dictatorship than with a democracy.
>
> Jeane J. Kirkpatrick, addressing Indians, 1981

Estrangement. How singular that two leading combatants of our time should, a decade apart, have reached for and grasped the same evocative term to describe Indian-American relations.

In 1972, when Indira Gandhi spoke of estrangement, Richard Nixon and Henry Kissinger had dispatched the aircraft carrier *Enterprise* on an exercise in nuclear gunboat diplomacy in the Indian Ocean. The revolution in Bangladesh and the consequent breakup of Pakistan had come at an inconvenient moment for Kissinger's China policy and for the Pakistani regime that was facilitating it.

In 1981, when Jeane Kirkpatrick proclaimed estrangement while leaving the New Delhi airport, she was already the spokesperson for a triumphalist new administration in Washington, representing a bellicose American political culture, predictably offended by Indian equivocation, most recently on Afghanistan.

But estrangement. The phenomenon goes far beyond personalities and Indian-American relations, and it transcends normal diplomatic vexations. In recent years, official America has periodically felt estranged from governments in every continent and region, from Tokyo to Brasilia, from Paris to Mexico City, from Ottawa to Bonn. Moreover, in the 1980s broad sectors of significant opinion inside other societies have felt increasingly estranged from official America. This is true in societies as traditionally friendly as democratic Britain and Scandinavia, not to mention Latin America, black Africa, and the Arab world, where skepticism toward U.S. policy is more ingrained.

Estrangement has begun to infiltrate transcultural attitudes themselves. Not all foreigners continue to distinguish between American policies they oppose and an American society that still merits esteem. And many Americans are quick to smear as anti-American any foreigners holding sentiments they happen to dislike.

Of course, estrangements need not last forever. Few things in politics do. In human affairs, separations can be superseded. Even genuine reconciliations have been known to occur. And among nations, official relations are often maintained despite estrangement.

Still, the very possibility of an estranged America would have seemed a contradiction in terms to the man who founded the Carnegie Endowment for International Peace. To Andrew Carnegie, a Scottish-born American superpatriot, America was inherently empathetic, exemplary, pacific, irresistible in its magnetism. A hundred years ago he dedicated his book, *Triumphant Democracy*, "To the Beloved Republic, with an intensity of admiration which the native-born citizen can neither feel nor understand":

> The weakest nation may rest secure, Canada on the north and Chili [sic] on the south, for the nature of a government of the people is to abjure conquest, never to molest, but to dwell in peace and loving neighborliness with all. . . . How many useless wars in the past would have been avoided had the republican methods prevailed! How many in the future would be prevented by its prompt adoption. . . . It is a proud record for the Democracy that the giant of the Western Continent is not feared, but is regarded with affection. . . . So, my fellow Republicans, the world is coming rapidly to your feet, the American Constitution is being more and more generally regarded as

the model for all new nations to adopt and for all old nations to strive for.

In 1910, when Carnegie established the Endowment, peace was in his mind largely a projection of American experience, ideals, methods, law, and political inspiration. He viewed the then-limited American connection with the world as uncomplicated, positive, and beneficent. The phenomenon of American estrangement would have been totally outside Carnegie's comprehension.

When the Endowment trustees in 1985 asked themselves how they might mark the foundation's seventy-fifth anniversary, they thought of many historic transformations that have occurred over the lifetime of the institution. The radically different assumptions, both at home and abroad, about America's relevance to the world and its engagement with the world, especially since the glory days of 1945, seemed an evocative point of departure, posing many questions:

Are we caught in a situation not of our own making, enmeshed in the webs and biases and disappointing reactions of others?

Are we estranged from them, or they from us, or are we mutually estranged in an inextricably interacting dynamic?

On our part, has familiarity with others bred contempt, or is it our lack of awareness about others that leads us into miscalculation and misdirection?

How much of our dilemma is our unconscious ethnocentrism, grating on the nerves of those whom we presume to lead or instruct?

How much is traceable to our economic and social failures at home, measured against our frequently self-righteous stance abroad?

Granted that the Reagan administration is achieving new depths of estrangement, was not the ground for it well seeded by the legacies of Johnson, Nixon, and Carter?

How much has estrangement proceeded from America's military misadventures and defeats of recent years?

How much of it reflects our reemerging know-nothingism, the atavistic stirrings on our political–cultural scene?

How much derives from our styles of presidential leadership and the consequent confusions these create abroad?

How much is simply American impatience with history, suspicion of diplomacy, and chagrin over the intractabilities of world events?

The Endowment decided to seek thoughtful essays on such

topics from a range of authors of established reputations and varying perspectives. For the challenging task of shaping a collection of essays a sensitive and talented editor was required. Fortunately, Sanford J. Ungar was available for this task. Formerly managing editor of *Foreign Policy* magazine and recently host of "All Things Considered" on National Public Radio, Ungar has produced a provocative and illuminating volume.

The Carnegie Endowment hopes that the insights in these essays will contribute constructively to the reappraisal that serious Americans ultimately must make of their national predicament as they confront the closing years of this tumultuous century. That predicament leaves us wondering how much to curse fate, blame others, or look inward. For, at the end of the day, we must also ask ourselves, with Cassius, whether the fault is really in our stars or in ourselves, that we are underlings.

Not, in the experience of the present writer, since the Harding era when we denied our enlightened self-interest and retreated from responsibility in our foreign relationships, while confessing to scandal and tawdry commercialism at home, has the world had such a poor opinion of us. American principles, which sometimes were characterized as naive but in general were respected as sincere and humane, now are freely called hypocritical and self-serving; the weight of American material and military power, looked to in the past as a mainstay of world stability, is now mistrusted and feared.

Hamilton Fish Armstrong, in his farewell article
as editor of *Foreign Affairs*, October 1972

Between Americans and Europeans—I say Europeans and not the French in particular—there is a gradual divorce. We no longer speak the same language. Americans are totally indifferent to our problems. For a long time we have been telling them: Your interest rates are too high, so we cannot invest. And if we do not invest, we cannot create employment. And now, on all these subjects we face this lack of understanding. . . . And when I say we, I do not mean we Frenchmen alone, but the Germans, the English, Italians, British Conservatives and French Socialists. There is a remarkable lack of understanding between Americans and Europeans at the moment. And this is very serious because it is a fact that the United States is the greatest country in the world and the United States is our principal ally. We must be able to talk to one another. We are farther apart in distance and in dimension. We must be able to talk to one another and at the moment we no longer do so.

Claude Cheysson, then French foreign minister, in an interview
with the television network *Antenne 2*, July 21, 1982

When one is a stranger to oneself then one is estranged from others too.

Anne Morrow Lindbergh, *Gift from the Sea*,
Chapter III, 1955

ESTRANGEMENT

1 The Roots of Estrangement

An Introduction

SANFORD J. UNGAR

Sanford J. Ungar, currently a senior associate at the Carnegie Endowment for International Peace, is a former managing editor of *Foreign Policy* magazine; from 1980 to 1983 he was host of "All Things Considered" and other programs on National Public Radio. He has also been a national staff writer for the *Washington Post* and Washington editor of the *Atlantic*, and is the author of several books, including *The Papers & The Papers: An Account of the Legal and Political Battle over the Pentagon Papers*, which won the George Polk Award. Mr. Ungar's latest book, *Africa: The People and Politics of an Emerging Continent*, was published by Simon & Schuster in 1985.

CERTAIN EVENTS in the life of a nation and a people are almost impossible to forget; they embody hopes and fears and other intense emotions, and eventually they become symbols of a mood or a moment in history.

Asked to identify such events from the 1940s, Americans would surely choose Pearl Harbor and Hiroshima, the Berlin airlift, and perhaps the founding of the United Nations. From the 1950s, they might select President Harry S Truman's decision to intervene in Korea, the Supreme Court's judgment outlawing school segregation, or the Soviet launch of the first Sputnik satellite. From the 1960s, they would certainly remember the assassination of President John F. Kennedy and could draw from a long list of other dramatic events: the Cuban missile crisis; the construction of the Berlin Wall; the Tonkin Gulf Resolution, the Tet offensive, and other dramatic incidents during the American involvement in Vietnam; the civil rights marches on Washington and from Selma to Montgomery, Alabama; the assassinations of Martin Luther King, Jr., and Robert F. Kennedy; and the chaos of the Democratic National Convention of 1968 in Chicago or some other stirring of the peace movement and the counterculture.

From the 1970s, the Watergate scandal and the resignation of President Richard M. Nixon dominate the national (and to some extent even the international) memory, but there are other developments from that decade—not unrelated to each other—which compete for recollection: the clash between the federal government and the media, especially over publication of the "top secret" Pentagon Papers; the dramatic American opening to China; the OPEC oil embargo; the fall of Saigon; and the Camp David peace accords between Israel and Egypt.

Yet there is still another event from the 1970s—from the very end of the decade—that was, in retrospect, the most important and far-reaching of all: the hostage crisis at the U.S. embassy in Tehran in which at first sixty-six, and ultimately fifty-two, Americans were held by militant Islamic fundamentalists as a protest against Washington's long-standing support for the Shah of Iran. That it began one year to the day before the U.S. presidential election of 1980 and ended just as Ronald Reagan was being sworn into office on January 20, 1981, makes it all the easier to mark the hostage crisis as a watershed event in recent American history. It touched raw nerves and provoked a frantic overreaction. In the end, it was the ultimate notification to the United States that its standing overseas had changed dramatically and would never be the same again. More even than the American disaster in Vietnam, the Iran hostage crisis represented the end of a postwar era of illusion for the United States; it remains a symbol of America's bewilderment and incapacitation.

The fall of the Shah earlier in 1979, following the return to Iran of the exiled Ayatollah Ruhollah Khomeini, was an event whose importance was at first overlooked and ignored by many Americans. But the connection with the Shah was the thickest and most complicated relationship the United States ever had in the Third World; it engaged major corporations, universities, foundations, the media, and other key American institutions. The Shah was the ultimate American client. His dynastic pretensions were indulged and his authoritarian habits overlooked, all in the name of friendship and strategic calculation. Despite clear signals of human rights violations and the suppression of dissent in Iran, President Jimmy Carter, during the Shah's visit to Washington in November 1977, said that Iran had "blossomed forth under enlightened leadership" and had become "a very stabilizing force in the world at large." The next month, spending New Year's Eve in Tehran, Carter proclaimed Iran "an island of stability in a turbulent corner of the world." In October 1978, after the Shah's regime had begun to crumble, Carter told Crown Prince Reza, the Shah's son, during a photo session in the Oval Office, that "our friendship and alliance with Iran is one of the important bases on which our entire foreign policy depends."[1] Although some of the Shah's behavior did not always seem so loyal to the United States—he was, for example, a leader of the "price hawks" in OPEC who drastically raised the cost of energy to America and to many more vulnerable nations around the world—he seemed to carry American prestige with him.

Even so, had the Shah gone quietly, only to be replaced by another willing U.S. surrogate, many Americans would hardly have noticed. But his fundamentalist Islamic successors were determined not to let that happen. And once the American hostages had been taken, the ghastly vision of hundreds of thousands of anti-American demonstrators, some flailing themselves with chains as they marched past the embassy and denounced "the Great Satan," became a disturbing and unforgettable image, fixed in the American psyche. It was literally a nightmare, brought by the miracle of technology into American living rooms every evening just in time to disturb the nation's sleep. (Indeed, ABC's "Nightline," a popular television news broadcast, is one of the enduring legacies of the hostage crisis; it grew out of that network's nightly special reports, entitled "America Held Hostage.") If television coverage of Vietnam had converted the American people into temporary pacifists, the video images from Iran turned them into a pack of snarling wolves, and not so temporarily.

The Iranian militants took advantage of satellite linkups to protest the evils of American-inspired modernization. They cut immediately to the quick of America's national ego. The hostage crisis —crude and random episode that it seemed to be—sparked an extraordinary surge of American patriotism and xenophobia, reminiscent of the reaction when the battleship *Maine* was blown up in Havana harbor in 1898 and the United States went straight to war with Spain. Politicians demanded revenge against Iran, and an angry public took its own feeble steps—reacting to their country's humiliation in songs, on T-shirts and bumper stickers, and, in scattered ugly incidents, by attacking Iranian students and others from the unfathomable Third World. The drama in Tehran quickly became the preeminent symbol of a world gone awry, where nothing seemed to be the way it used to be, or (more accurately) the way Americans had come, since World War II, to expect it to be. Precisely because there was no way, place, or method to go to war or otherwise obtain vengeance without disastrous side effects, the nation felt all the more frustrated and betrayed.

Once the Iran hostage crisis had begun, it was easy enough to indulge in a retroactive blaming process, to go back and reinterpret many events and trends since World War II as signs that the world was going bad, if not mad. Ho Chi Minh, Charles de Gaulle, Leonid Brezhnev, Fidel Castro, Indira Gandhi, Muammar al-Qaddafi—they and a cast of a thousand others could now be seen as co-conspirators in the effort to make the earth a precarious place for the United States and its well-intentioned leaders to live. Suddenly America had to fend for itself and could count on no one else. It was bad enough that the United Nations, the very epitome of postwar American idealism, had turned into what one U.S. representative there, Daniel Patrick Moynihan, called "a dangerous place."[2] But worse, the allies acted unreliably, the beneficiaries of American largesse seemed ungrateful, and even adversaries did not behave predictably. Everywhere the American people turned they found new evidence not only of grave problems but also of Washington's apparent inability to solve them.

If events were examined closely and calmly, outside the climate of hysteria that so often prevailed in Washington, it was possible to see that America's changing circumstances were part of a general shift in international relationships. Other nations and people, some of them having only recently emerged from the experience of colonialism, were asserting more than symbolic independence and using whatever power they could muster to seek greater control over their own lives and national destinies. But that was not how things looked and felt from the perspective of an embattled, weary America. And so in many instances, when other actors on the international scene discovered the dramatic effect their rhetoric could have on the United States—the severe aggravation they could cause with simple gestures—they escalated it. The greater the tendency of Americans, including presidents and secretaries of state, to interpret an evolving world as a deteriorating world, the more the United States seemed to be losing control and influence. There was an element of self-fulfilling prophecy to it all.

When the Soviet Union invaded Afghanistan less than two months after the hostages had been taken in Tehran, Carter tried a smorgasbord of sanctions, not a single one of which was effective. Indeed, Carter's ultimate means of expressing his distaste for Soviet actions, the American-led boycott of the Moscow Olympics in 1980 (on which the United States spent an enormous amount of diplomatic and political capital), did more harm to the U.S. image than it did to Soviet prestige. As almost anyone could have predicted at the time, it produced a Soviet-led counterboycott of the Los Angeles Olympics four years later. In the end, the United States found itself blamed for politicizing international sports.

West Europeans seemed incapable of sharing American outrage over Soviet behavior or of being persuaded to fall into line. They refused to institute severe sanctions over the Afghanistan affair or, even later, over the declaration of martial law in Poland. On the contrary, in the midst of this critical chain of events, the Europeans signed a contract with the Soviet Union to construct a natural-gas pipeline and granted Moscow credit on favorable terms.

In Lebanon during the early 1980s, the United States proved utterly unable to provide anything but targets for terrorists. The Reagan administration's brief foray into on-the-ground Middle East peacekeeping there turned into a disaster, costing the lives of hundreds of American servicemen but in no way advancing the cause of local or regional reconciliation. In Nicaragua, the Central Intelligence Agency and its "contras" were incapable of overthrowing the Marxist government of the Sandinistas, or even, in the president's own inelegant term, of making them "cry uncle." And in the case of the tiny Caribbean island of Grenada, where the Pentagon did manage to oust a small Cuban garrison and the radical fragment of a left-wing regime—scoring what passed, at least among most Grenadians and Americans, for a great success—the State Department could not find many friends or allies to pat America on the back. The vote in the U.N. General Assembly on a resolution "deeply deploring" the American move into Grenada was 108 to 9, with 27 abstentions. The only supportive votes came from a few other Caribbean states, El Salvador, and Israel. Many countries regarded as stalwart supporters of America voted to condemn the invasion, and the best the United States could get from Britain, West Germany, Japan, and Canada were abstentions.

It was hard to imagine what other insults might be devised to wound the American ego, but a usually compliant Pacific ally, New Zealand, found one early in 1985. The newly elected Labor party government there declared that it would not permit a U.S. Navy destroyer to make a port call unless it received assurances that the ship was not carrying nuclear weapons. Such assurances being against long-standing American military policy and practice, the Reagan administration fell into a bitter public quarrel with a nation that had sent troops to fight alongside Americans not only in World War II but also in Korea and Vietnam. It would not have been an easy dilemma for even the most skillful negotiators to solve, but the Reagan administration made matters worse by overreacting to the heartfelt fears of New Zealanders, who had had their atmosphere poisoned over the years by the fallout from French nuclear tests in the South Pacific and thus felt a more than routine abhorrence of nuclear weapons. Hinting at trade sanctions, the United States came off looking like a bully frustrated by his inability to get everyone else to play his game by his own rules. What might have remained a minor incident was blown far out of proportion, because it touched another sensitive nerve among Americans. Before long, rather than searching quietly for a compromise, Washington virtually nullified ANZUS, the mutual security pact the United States had signed with Australia and New Zealand in 1951. A mouse had roared, and the startled American lion characteristically responded by trying to punish it for disturbing the peace.

The United States emerged from World War II with a unique position and an unusual role in the world. Rarely in history had a single nation enjoyed so much prerogative and influence. Alone among the major actors on the international scene, America had survived a nearly global conflict intact and, as a result, perceived itself to be capable of enjoying political and economic ascendancy over its allies and adversaries alike. There was no immediate evidence that this kind of domination would not be sustainable. The American public demanded peace and power, and the country's leadership, regardless of party, felt confident that it could provide both without straining the domestic political system or the new world structure that the United States had guided into existence.

Indeed, U.S. foreign policy in the postwar period was conducted on the basis of a simple vision of the world and of the American role in it. The self-image of the United States was that of a keeper of world peace. As the primary sponsor of the new international order drawn up in San Francisco (the United Nations) and at Bretton Woods (the World Bank and International Monetary Fund), the American government believed that its role in the allied victory in Europe and the Pacific brought with it broad responsibilities for the indefinite future and assurances of cooperation from every quarter. Whether in the context of the United Nations, the North Atlantic Treaty Organization, or on its own, the United States wanted to put out brushfires that threatened international calm. Because the assignment seemingly had universal endorsement, carrying it out was not expected to provoke serious dispute at home or dissatisfaction abroad.

One way that America initially sought to keep the peace was as leader of a broad new Western alliance and, more ambitiously, of an entire "free world" arrayed against communism and other menacing forces. The credentials for membership in this free world were sometimes confused and often required stretching the language to accommodate dictators and tyrants. Moreover, the United States, self-confident and cocksure, saw no necessity for occasional reviews of its performance as leader; in the beginning, at least, few other countries were in a position to criticize or complain.

America's confidence was based in part on the assumption that it could be the dominant party in the new but quickly expanding rivalry between the postwar superpowers. This competition was aggressive, sometimes openly angry and highly dangerous, but the United States believed—and, again, saw little convincing evidence to the contrary—that it enjoyed a technical and, even more important, a moral superiority over the Soviet Union. From the perspective of the late 1940s, with the U.S. nuclear monopoly intact, these assumptions seemed entirely reasonable. Many U.S. and European policymakers and analysts even regarded the emerging superpower standoff as an ultimately stabilizing element in world affairs. With the globe neatly divided into two camps and no one trying to occupy a confusing middle ground, it would be clear where everyone stood. The United States would face no surprises, and it could manage any new crises that did arise. Or so it seemed.

The America of the late 1940s also saw itself as a role model—a political, philosophical, economic, and cultural hero to the developing societies that would soon become autonomous, sovereign countries. By all available evidence and, the West believed, by any rational process of reasoning, these societies would be inspired not by Marxism but by Franklin D. Roosevelt's emblematic Four Freedoms and the vision of equality contained in the Atlantic Charter. The concept of a "Third World" with pretensions of going its own way was almost impossible to imagine, especially for the United States, a country whose example now seemed to have something approaching a divine force behind it.

Consistent with a commitment that went back to the early years of the nineteenth century, the United States intended to act more than ever in the postwar period as the protector and defender of the Western Hemisphere from incursions by outside forces and ideas. There had been occasional upheavals in Latin America before World War II (Augusto Sandino's uprising in Nicaragua, for example), but Washington had usually been able to restore order by marching right in. With the new levels of American power and influence, it seemed as if this would be easier than ever to do. The *Pax Americana* would be extended elsewhere, too—notably in Vietnam and the Persian Gulf—but circumstances in its own hemisphere often had a profound effect on America's policy and behavior in other parts of the world. The rise of Fidel Castro in Cuba and his affiliation with the Soviet Union, a subversive event that took place only ninety miles from the tip of Florida, thus intensified the American determination to prevent similar unsettling developments elsewhere, near or far.

It is by now a commonplace that the comfortable, or at least familiar, postwar world is in its twilight, but the implications of the change are still unfolding and only beginning to be understood. The contrariness of a former wartime partner, Charles de Gaulle, in the 1960s sent the Western alliance into a process of disintegration. Misunderstanding became the order of the day, and although the United States and the other European allies managed to coordinate on military matters without France, their political relations were often full of recrimination. There were quarrels over trade and other elements of international economic policy. Certainly the Europeans, having belatedly learned the lessons of nationalism, could not understand what the United States was trying to do in Vietnam; later they refused to defer to Washington even in Central America. In one crisis after another, the United States saw itself as the solution, but in nearly every case a significant number of other nations and their leaders began to see the United States as part of the problem. The term *free world* lost its meaning, to the point of becoming an object of mockery and derision, and America's capacity to lead what was left of that world was called seriously into question.

The assumption that the U.S.-Soviet rivalry imposed a kind of stability on an unruly world withered quickly. The Soviet Union proved unwilling from the start to play its assigned role of junior partner in the postwar superpower competition. Once it had its own nuclear arsenal, there was no reason for it to do so. For a time American rhetoric seemed more believable, and U.S. nuclear power more benign, than their Soviet counterparts; but even America's moral edge—its special claim to credibility—faded over the years, especially in the view of the growing bloc of "nonaligned" countries, which blamed both sides for the world's parlous state. The concept of mutual deterrence, an important element in the postwar balance of power, became less reassuring and reliable, and ever more fragile. Increasingly, the U.S.-Soviet contest looked like a destructive, rather than a constructive, organizing force; it demanded a vast share of both societies' resources and left much of the rest of the world quivering at the prospect of nuclear catastrophe. Meanwhile, the recurring issue of whether America could retain, or reestablish, the superiority it had enjoyed in the immediate postwar period provoked a caustic domestic political debate in the United States and often contorted not only superpower relations but also the rest of U.S. foreign policy.

Far from fulfilling America's naive hopes and dreams and looking to its leadership, the Third World became a caldron of unrest and despair. The United States often experienced the shock of being regarded as an ally and extension of colonialism, rather than as a society that offered inspiration for future development. In the United Nations and other international forums, American diplomats came gradually to be subjected to a torrent of abuse and accusations that tried the nation's patience and provoked its temper. Overseas they were frequently threatened by acts of terror and violence, and some lived with a daily fear of assassination.

The most painful events were those closest to home. In Latin America, once Castro had come to power, the tide seemed to run inevitably against perceived American interests, and credible surrogates—not to mention simple friends—became ever harder to find. Instead of the just and intrepid "Good Neighbor," the United States was increasingly portrayed as a grotesque caricature, the evil purveyor of "Yanqui imperialism."

There and elsewhere, American policymakers tried to promote the ideal of stable change managed by moderate forces. But rarely did that work, and when the extremes clashed, the United States generally found itself, wittingly or unwittingly, in alliance with conservative, often repressive forces opposed to change. The persistent tendency by American administrations of both parties to cling to symbolic leaders on their way out—the Shah, Anastasio Somoza in Nicaragua, Haile Selassie in Ethiopia, and Ferdinand Marcos in the Philippines, to name a few—reinforced the impression. In a great range of circumstances, from Vietnam in the 1960s to southern Africa in the 1980s, the United States became the object of vilification and mistrust by the majority of the people. The great nation that wanted to be a peacemaker somehow seemed more a party to than a preventer of conflict. And conflict was becoming more likely, in more places, all the time. As the proliferation of nuclear-weapons-making capability went virtually unchecked, notwithstanding American moral appeals and the Nuclear Non-proliferation Treaty, the gravity of potential conflicts grew, and the prospect of disorder deepened.

American politics and U.S. foreign policy adapted awkwardly

to the breakdown of the world the United States had tried to create and lead. For a brief period John F. Kennedy seemed to offer America and the rest of the world fulfillment of the frayed postwar vision after all, but beginning with Lyndon B. Johnson, the succession of men who came into the White House were treated as objects of international puzzlement rather than admiration. Courses in American Studies overseas became a combination of psychohistory and sociogeography. The impasse in Vietnam, the U.S. domestic upheaval growing out of it, and the civil rights movement, combined later with the extraordinary collection of events known as Watergate, led many Americans to cease their perennial lectures to the people of other countries about consensus or corruption. Other Americans kept lecturing, but few people in the intended audience were listening. Peace and political legitimacy were becoming hard enough to attain at home; America hardly had the moral qualifications or the physical capacity to establish peace and project its power around the world anymore.

With the U.S. economy threatened by its own decline in productivity, the growing assertiveness of oil producers, and competition from other rapidly expanding industrial economies, the American success story became a less obvious model to admire or emulate. Indeed, a German chancellor and a French president who knew something about economics were baffled by American presidents who did not. Meanwhile, much of the domestic discussion of international issues deteriorated into an examination and dissection of the "post-Vietnam syndrome," as if some failure of will there had not only constituted a stab in the back in Southeast Asia but had also induced a crippling caution that now prevented the United States from taking effective action anywhere. Some people began to search for another place—perhaps Central America—where a new group of tough American leaders could "do it right." Nearly everyone craved simplicity in foreign affairs. It was as if a cascade of unhappy events—the climax being the prolonged hostage crisis in Iran—had sent the United States back into a parochial, unsophisticated pre–World War II attitude toward an ever more complex world.

The United States is estranged from that complex world—separate, aloof, more alone than even the most cynical or pessimistic observers might have predicted in the heyday of American postwar power. The symptoms are obvious everywhere: in trade friction with Japan; in the antinuclear, anti-American sentiment infusing European political movements; in some countries in Central America, where only the most conservative, most narrow-based elements will align themselves openly with U.S. policy; and in South Africa, where recognized black leaders have become cautious about meeting with even liberal Americans who agree with them and support their cause. Although it often emerges in the form of bravado or contempt for others, the unease is also deeply felt inside the United States. It is a phenomenon that extends across party and ideological lines.

Part of what makes today's world so uncomfortable for Americans is the perfectly understandable tendency to reject a new reality, especially strong in a society where self-confidence is part of the national ideology. "Americans feel unqualifiedly that this is the best country in the world," the pollster Daniel Yankelovich has observed. They are taught to believe that from their earliest days in school, and any politician who disputes this fundamental truth (as Jimmy Carter seemed to do toward the end of his term in the White House, when he worried aloud about the American "malaise") is severely punished for his heresy. "The dark side of this attitude is that we don't believe we can be wrong," Yankelovich went on; "we're not looking at the world from anybody else's point of view."[3]

The United States has provided frequent reminders in recent years that it sees the world through its own unique, self-interested prism. On three issues that became international symbols, America's isolation was particularly stark: In May 1981, at a meeting of the World Health Organization in Geneva, the Reagan administration cast the sole dissenting vote against the adoption of an international code of ethics curtailing the promotion of baby formula. (The issue had arisen because of concern that the aggressive marketing of formula in Third World countries, where it was often mixed with contaminated water and fed in unsterilized bottles, was causing an increase in infant deaths.) In April 1982 the American delegation to the U.N. Law-of-the-Sea Conference cast one of only four votes against a draft treaty, because of objections to the provision establishing a global authority to regulate seabed mining. At the end of 1984, although 138 other nations and 21 international organizations had already signed, the United States (influencing Britain, West Germany, and 9 other countries, with varying degrees of reluctance, to go along) let the deadline for signatures pass, effectively eliminating itself from any role in the cooperative international management of ocean resources. At the U.N.-sponsored International Conference on Population in Mexico City in August 1984, the United States found itself virtually alone against the rest of the world in arguing for the adoption of a fundamentalist anti-abortion policy and in attempting to insert language promoting free-enterprise economic policies into resolutions that came before the conference.

There was a time, in the early postwar period, when a preoccupation with America's well-being was not necessarily regarded as inconsistent with a genuine concern for and generous interest in the rest of the world. Christoph Bertram, political editor of the West German weekly *Die Zeit* and former director of the International Institute for Strategic Studies in London, recalls an era "when the United States was a superpower in the true sense: confident not only in its strength but also in its ability to build, together with others, a world of shared duties and rights, and to be ready to carry the major burden in this enterprise."

That was an era when Marshall McLuhan's vision of the "global village" was just dawning, when it was still reasonable to dream of a world community in which geographical and social distances were rendered meaningless by methods of communication that promoted understanding—restoring, McLuhan believed, something of our species' original sense of being part of a village or tribal unit. The importance of physical distances, and of some of the social ones, has been reduced even further in recent decades, but the concept of a global village now seems naive and the notion of the United States taking responsibility for it, in the current domestic and international political climate, nothing short of ridiculous. "America has given up the traditions it established after World War II," says Bertram, "without regrets but with a sigh of relief." Rather than promoting a new "international order that promises cooperation and stability for the 1990s, America prefers to pursue its own interests alone," he says. Bertram concludes that "we are now back in [a world] where the strong do what they want and the weak suffer what they must, where superpower, once again, is a measure of military strength alone."[4]

There are many competing diagnoses of the problem, of course. Conservatives tend to blame America's lack of resolve. For Jeane J. Kirkpatrick, U.S. ambassador to the United Nations during the first Reagan administration and a leading proponent of the new

American assertiveness, the critical change occurred during the late 1960s and the 1970s, when the United States suffered from "an attitude of defeatism, self-doubt, self-denigration, and self-delusion." One of Kirkpatrick's predecessors, Moynihan, pointed to a pervasive American sense of guilt and "a failure of nerve" within the U.S. foreign policy elite. "Inhibitions had arisen in the American polity," Moynihan observed, "which prevented us from responding to ideological conflict." Michael Ledeen, once an aide to Reagan's first secretary of state, Alexander M. Haig, Jr., complains that in recent years American leaders "have tended to adopt moralistic abstractions instead of realistic appraisals of the world, and we have rarely suited our diplomatic, economic, and military means to our national objectives." The problem is compounded, according to Ledeen, by the fact that "fashionably correct positions on many issues are defined by a group of journalists who are themselves ignorant of the dynamics of international affairs and who are themselves explicitly engaged in political activity with strong professional investments in certain outcomes."[5]

Liberals, on the other hand, think America has closed its eyes and ears to many developments in the postwar world. Jonathan Kwitny, who has traveled the globe for the *Wall Street Journal,* believes that in the Third World

> we continue to press not our system, which encourages free choice, but some convoluted notion of our system, which imposes *our* choice. We insist on imposing solutions to particular problems involving foreign people. They are asked to live by our choices, when they often don't want or even understand them.

U.S. foreign policy is above all simplistic, according to T. D. Allman: "The most intractable myth of American nationalism is that there are intrinsically good guys and bad guys." And three analysts of the American foreign policy process—I. M. Destler, Leslie H. Gelb, and Anthony Lake—contend that the United States has

> lost a coherent sense . . . of national interests, the enduring purposes of policy that flow from values, geography, and our place in the hierarchy of world power. In almost all other nations, it takes a revolution to redefine these basic purposes. For the last two decades in the United States, it has required only a Presidential election or the prospect of one.[6]

For Fouad Ajami, a scholar of the Third World who was born in Lebanon and now teaches in the United States, the issues are more complex, involving the export of ideas by the United States that are somehow unrecognizable when they are espoused by others. "America has formed so many of its rivals," Ajami has written:

> The images we do not like in distant societies are often reflections of ourselves. America has held up before older societies a revolutionary message of social change and political equality; every now and then we ride into storms that we helped stir up. When we understand this, we will no longer imagine "others" as men of dark sensibilities in thrall to frightening forces. We will also begin to understand the deep roots of America's presence, as well as the ambivalence with which it is greeted by men who hector us in metaphors at once familiar and threatening, in places that invite us in and then reject us.[7]

Indeed, however it is analyzed and explained, there is in America's estrangement an implication of prior attachment and affection, a sense that at some other moment in the recent past the United States and the rest of the world had more in common or understood each other better. It is as if a once very popular member of a family suddenly found himself at odds with all his relatives; after such a relationship has begun to sour, it is difficult to stop the downward spiral. And it is virtually impossible to be estranged from someone unknown or little known; only because of an established familiarity is a relationship vulnerable to deterioration.

But an argument over familiarity—or a lack thereof—lies at the heart of America's uncomfortable estrangement. The leaders and the intellectuals of many other countries believe that they have been asked to appreciate American processes and points of view, and even to imitate them, without an equivalent effort being made in the opposite direction. The result is an astonishing asymmetry: the minutiae of American politics and society are followed compulsively by others as a critical component of life in the real world, while the U.S. government and individual American citizens seem to show an ever declining knowledge of and interest in everyone else. Whether out of discomfort, indifference, or contempt, Americans—even those holding high positions—often laugh at the hard-to-pronounce names of foreign capitals or statesmen while expecting others to treat American place and family names with reverence. But is Ouagadougou (the capital of Burkina Faso, formerly Upper Volta, in West Africa) really any more exotic a name than Massachusetts or Tallahassee? And is it any more eccentric for people from Spanish cultures to put their mother's maiden name after their surname than for Americans to insist on using middle initials?

The ignorance of Americans about international affairs is legendary, and it has been growing worse. Even on matters of current controversy—issues that are theoretically decisive in U.S. presidential and congressional elections—the public is remarkably unaware. Asked merely to tell the location of El Salvador in a survey sponsored by CBS News and the *New York Times*, only 25 percent of the sample got it right. When ABC and the *Washington Post* asked people to say which two nations had been involved in the SALT talks, only 37 percent answered correctly; 13 percent gave a wrong answer and 50 percent said they had no idea. In response to other questions, fewer than half of the people surveyed knew whether it was the United States or the Soviet Union that belonged to the NATO alliance or which countries had participated in the Camp David peace talks sponsored by the United States.[8]

Schoolchildren show little promise of improving on their parents' records. One study published in the late 1970s showed that more than 40 percent of American high school seniors surveyed could not locate Egypt on a map, and more than 20 percent had no idea where to find France or China. The President's Commission on Foreign Language and International Studies, reporting in November 1979, found a striking decline in the study of languages at the high school, college, and graduate levels. It warned that the network of institutions and special programs in international studies, "built by foundations, the government, and the universities themselves in a period of great scholarly advances after World War II," was now "in danger of imminent collapse."[9] The commission recommended a major effort to increase federal support and other funding for such centers, but since then the Reagan administration has proposed, in several successive fiscal years, reducing that item in the Department of Education's budget to zero. (Congress repeatedly resisted that proposal and insisted on funding regional centers for language and area studies, but education organizations fear that congressional resistance will soon disappear.)

The American government is notorious for sending diplomats overseas with little or no knowledge of the cultures, let alone the languages, of the countries where they will serve; among the U.S. embassy staff in Iran who became hostages, for example, only a handful were fluent in Farsi, the country's most important language. Political appointees are often selected for certain ambassadorial posts, especially in Europe and Africa, without any regard for the nominees' qualifications.

Given these educational trends and federal policies, it is little wonder that, as political scientist Howard J. Wiarda has noted, at the root of U.S. foreign policy dilemmas in the Third World lie "a deeply ingrained American ethnocentrism, an inability to understand the Third World on its own terms, an insistence on viewing it through the lenses of our own Western experience, and the condescending and patronizing attitudes that such ethnocentrism implies."[10] Such neglect and narrowness is bound to have lasting consequences for America.

Still, America's estrangement is not without its paradoxes.

Even as U.S. political and economic influence comes in for growing criticism and often virulent attack, American cultural symbols seem to flourish almost everywhere. For years, Americans traveling to the Soviet Union were advised to take along blue jeans, chewing gum, and ballpoint pens if they really wanted to have doors opened to them. European leftists sitting in cafés and complaining about the United States are as likely as not to be drinking Coca-Cola, and even in the tensest moments of Nicaraguan-American relations—when U.S.-supported "contras" are attacking, the Sandinistas' press censorship is at its peak, and the government is sounding alarms about an imminent U.S. invasion—movie theaters in Managua are likely to be featuring a festival of American films. McDonald's and other American-franchised fast-food outlets have become a permanent feature of life in Japan, as have other aspects of U.S. popular culture. Indeed, after World War II, during the American occupation of Japan, General Douglas MacArthur succeeded in shaping various aspects of public and private life there (including, for example, television networks) according to the American model; there have been complaints about the loss of Japanese traditions, but few serious efforts to do anything about it.

Jazz is virtually a lingua franca in the Third World, and in many places, if the Voice of America has taken hold, it is primarily because of its broadcasts of American popular music. American television serials such as "Dallas" or, in their time, "Gunsmoke" and "Kojak" have been dubbed and subtitled in so many languages and sent around the world so many times as to inspire local fan clubs and imitations. Professionals and intellectuals in many developing countries where the local press is tightly controlled wait every week for the arrival of American news magazines, so they can learn about developments on the international scene (and even in their own nations), and in other places where those magazines cannot be sold openly, copies fetch high prices on the black market. For much of the twentieth century Americans and Americanophiles had dreamed of the "Americanization" of the world,[11] but by the 1960s and 1970s the dream had, in certain respects, become a reality. And the Soviet system, for all of the American warnings about its spread, had offered no alternative, attractive models.

So thorough is the saturation of some parts of the developing world with American culture that truly revolutionary forces—as distinct from mainstream political opponents of a regime in power —often find they must attack symbols of that culture before they can be taken seriously. Thus, when the Shining Path guerrillas in Peru wanted to take their antigovernment campaign from the countryside to the capital city of Lima, one of their first targets was a chain of Kentucky Fried Chicken outlets.

Another paradox is that the American example of a good life achieved through material consumption and technological progress, while initially admired and envied, has ultimately led to alienation rather than appreciation. It was long assumed by U.S. officials, business interests, and journalists that modernization was the dream of the poor and the oppressed around the world. It was as if they expected the American dream to take hold so thoroughly that before long people everywhere would aspire to live in suburban homes with modern conveniences—to act out the lives they glimpsed in American television series or in advertisements for American products in American publications. But that, of course, was far from the truth. Most of the citizens of developing societies were, in the first place, concerned with far more basic problems, such as feeding themselves and keeping their children alive from one season to the next. And for those few members of the elite who did have access to Western ways and advanced education, the preservation of traditional values often became a more important consideration than many outside experts or aid donors imagined.

As Wiarda pointed out, the United States has been a prime sufferer from Eurocentrism, a "belief that the developmental experience of Western Europe and North America can be repeated, even imitated (albeit belatedly) in the Third World." Thus, Americans have expected economic, political, and social change to follow the Euro-American model, especially in countries of particular interest to the West, and they may be baffled when Islamic, Confucian, Buddhist and other traditions or sensibilities assert themselves. Iran is the classic example of a client state where American expectations dissolved as the domestic political situation went haywire. There the encounter with Western culture did not at all lead to an American-style pluralistic and secular society, but rather to a harsh, theocratic, ultimately totalitarian regime.[12] And the establishment of that regime was greatly facilitated by the easy access of ordinary people to high-tech implements. The Ayatollah Khomeini's subversive speeches against the Shah received their widest circulation on cassette tapes that were smuggled into the country and then heard on miniature tape players (manufactured in Japan, perhaps, but inspired by American habits).

Probably the most striking paradox—and the one that may allow Americans to overlook or dismiss their country's estrangement from the rest of the world—is that while the United States is often the target of international criticism, it nonetheless remains a mecca for immigrants. In fiscal year 1984 alone, some 544,000 people emigrated legally from their home countries to America, and the Immigration and Naturalization Service apprehended more than 1.2 million others who had come illegally. It is indisputable that however unpopular the United States may be on the international political scene, a vast percentage of humanity still regards America as the promised land where economic opportunity and a liberal political system make it possible to dream of a better life. This overwhelming attraction fortifies the self-congratulatory impulse of many Americans and helps them reject national introspection and self-criticism in favor of defiance, anger, and a new nationalism.

There is a natural temptation, especially during difficult periods in their nation's life, for Americans to try to withdraw, literally or figuratively, from international involvement—to rely on their country's geographical isolation and self-sufficiency and try to live apart from the world's turmoil, as it once seemed possible to do. But it is no longer feasible to reach for such an alternative. Georgi A. Arbatov, director of the Institute of U.S. and Canadian Studies in Moscow, has observed that

> the situation has changed, and America finds herself not only at a rough parity with the USSR militarily but absolutely equal with other

> countries in terms of her own vulnerability to a holocaust should a war break out. This is a new situation for Americans. It is undoubtedly not easy to get used to it, not easy to get along with it.[13]

Indeed, separation—one logical result of any process of estrangement—is not an option in today's world, especially not for an open and influential society like America, still preoccupied with peace and power. An isolated, backward nation like Albania can, for quite some time, withdraw into itself and keep foreign influences from crossing its borders. A xenophobic culture like Burma can restrict visitors to short stays. Even the world's most populous country, China, could undergo a "cultural revolution," turning into itelf at the expense of impoverishing itself further and delaying its entry into modern life. What America must do now is uniquely complex and difficult: It must rethink its role in the world, and it must begin by recognizing that old myths of invulnerability and invincibility are no longer sustainable. As Ronald Steel has written, "the United States is still incontestably the richest and strongest nation in the world. . . . It can do many awesome things: move mountains, topple governments, confer riches, and inflict horrors." But that power is now finite and greatly circumscribed. "The reason that [America] cannot always call the tune is to be found not in the decline of American will," Steel observes, "but in the recovery of the world, a recovery we did so much to make possible. Now we have to learn to live with that recovery."[14]

One way to adapt to the world's recovery—and to its evolution into a seemingly less congenial place for America—is to examine carefully the events since World War II and to seek a new understanding of the practical and philosophical dilemmas that have arisen for the United States during that time. That is the purpose of the essays in this volume. Some deal with particular moments in America's postwar experience, others with broad themes and policy issues that emerge from the past four decades. But their common goal is to evaluate America's position in the world and to illuminate the origins of current problems.

Robert Dallek describes the immediate postwar setting, when the United States took the lead in constructing the new world system that would later become so perplexing. He explains how American values and expectations inspired, and yet eventually undermined, the new international institutions, and how President Harry Truman's attitudes and actions set the stage for later policies.

Truman's fateful decision to involve the United States in the

Korean War is at the heart of Robert Donovan's essay. He illuminates the domestic and international pressure leading to that protracted, bloody, and ultimately frustrating conflict, and draws lessons from it that make later U.S. foreign policy tragedies more comprehensible.

Donald McHenry examines the ambivalent American attitude toward the rise of nationalism in the postwar period—the apparent contradiction between the U.S. roles as an inspiration for people seeking self-determination, on the one hand, and as the primary status quo power questioning the course of decolonization, on the other. In this dilemma McHenry finds the source of later U.S. difficulties in dealing with the Third World.

J. Bryan Hehir considers the momentous events of the 1960s, when nuclear strategy and the politics and ethics of American power underwent a profound transformation. In Hehir's view, the perspectives of that decade influence the American debate over international affairs in the 1980s, especially as the United States looks back and tries to reinterpret the causes and consequences of its experience in Southeast Asia.

The 1960s and 1970s are also the focus of Godfrey Hodgson, who argues that domestic developments in the United States, including the civil rights movement and the Watergate scandal, had a major effect on American relations with the rest of the world. Tracing the disruption of the Western alliance during Charles de Gaulle's presidency of France and the influence of the Vietnam War on America's international standing, Hodgson portrays the decline of U.S. confidence in its own capacity to act forcefully.

Lester Thurow reviews the dynamic, and sometimes misunderstood, relationship between economic and political-military power in the postwar era. He examines the anomalous situation of the United States as the leader of an alliance that risks falling behind its partners economically, and he warns against a politically explosive drift toward protectionism and other policies that would hinder international economic cooperation.

Taking up the asymmetry of American relations with the Third World, Ali Mazrui complains of the inability of the United States to listen to messages from those who are less powerful. He calls attention particularly to the distinct challenges posed for America by Islam and Marxism and warns against confusing the two.

Searching for the roots of current American foreign policy in the early years of American independence, Philip Geyelin finds a disinclination to become involved in extended, demanding overseas affairs that was first enunciated by John Quincy Adams. The conflict between this underlying disinclination and the desires of recent presidents and secretaries of state for international leadership, he says, is responsible for the image of incoherence and disarray that afflicts U.S. initiatives today.

James Chace, in contrast, detects in American history a strong propensity to intervene with force in the affairs of other nations, primarily as a consequence of a persistent national sense of vulnerability. Focusing on the actions of the Reagan administration, especially in Central America and the Middle East, Chace says that it is time to distinguish between America's vital interests and its less crucial general interests.

Frances FitzGerald, broadly reviewing the intellectual history of America's attitudes toward the world, discovers a messianic vision with roots in nineteenth-century Protestant theology. She asserts that the United States, if it seeks to cope more effectively with international challenges, must begin to see the world for what it is rather than what it would like the world to be.

In his concluding essay, Richard Ullman considers the phenomenon of estrangement from a conceptual perspective, finding that the problem exists on several levels and seriously complicates international relations. The preoccupation with U.S.-Soviet ties, Ullman argues, distorts other areas of American foreign policy and adds, in particular, to the strains with Europe and the Third World. Unless this competition can be rendered less threatening, he says, international tensions will grow and the United States will find itself increasingly alienated from its own natural constituency around the world.

2 The Postwar World

Made in the USA

ROBERT DALLEK

Robert Dallek's writings on American history and politics include *Franklin Roosevelt and American Foreign Policy, 1932–1945*, for which he received the 1980 Bancroft Prize for history; *The American Style of Foreign Policy*, a reinterpretation of America's international role; and the recently published study, *Ronald Reagan: The Politics of Symbolism.* Mr. Dallek is professor of history at the University of Califorina at Los Angeles and a research associate at the Southern California Psychoanalytic Institute.

AMERICANS HAVE ALWAYS BEEN at odds with the outside world. And not without reason. Power struggles, wars, authoritarian governments, class divisions, extremes of wealth and poverty, and oppressed minorities were antithetical to American institutions and professed ideals; they made isolation from overseas affairs an attractive option and a fixed principle of foreign policy. In the era before high-speed air travel and military rockets, vast oceans on the east and west and weak neighbors to the north and south allowed the United States to shun political and military commitments abroad.

When the nation did reach beyond its borders in the period before 1945, its overseas involvements were cast as serving not national gain but larger moral ends. The emphasis in the War of 1812, the Mexican War of 1846–47, the Spanish-American War of 1898, and World War I was not on material advantages to be won in the fighting but on neutral rights, manifest destiny, freedom for oppressed colonial peoples, and a "war to end all wars." If European and Latin American governments complained of Yankee aggressiveness in these struggles, they did not make much of an impression on most Americans, who saw only an idealistic zeal for the betterment of the world behind their country's actions abroad.

American commitments to international well-being prior to World War II were more apparent than real. But in that remote, shadowy arena of the outside world, most Americans valued appearances more than realities. Unaffected by the frustrations and defeats that other nations had encountered in world affairs, Americans generally viewed external matters not in terms of what was, but of what they wished to see. U.S. actions abroad were less a response to external realities than a unilateral attempt, on the one hand, to advance narrow interests and, on the other, to confirm national values and worth. Yet, whether pursuing substantive, material goals or symbolic, psychological needs, Americans generally described their actions as cooperative gestures promoting world peace. The consequence was not genuine progress toward a more perfect world, but feelings of alienation abroad from a country that spoke one way and behaved another.

For foreign leaders who were accustomed to acts of cynicism in international relations, it was not simply the discrepancy between what was said and done that created tensions with the United States, but also the righteous, universalist rhetoric masking parochial acts and needs. When, for example, at the turn of the century, the United States preached the rule of law as it took control of the Panama Canal Zone, advanced the doctrine of the Open Door in East Asia, and sought to close out foreign influence in the Western Hemisphere through the Monroe Doctrine and the Roosevelt Corollary of 1904, Latin American and Japanese governments complained that the United States was applying a double standard to the conduct of international affairs. But they found little means to translate their objections into action, and so the protests barely registered on American minds. Down to the 1940s, then, the United States met little effective opposition to its will abroad, leaving it free to describe even its most selfish acts as altruistic steps in behalf of universal law and order.

Circumstances at the close of World War II reinforced American impulses to ignore external realities and to alienate foreign governments and peoples. In 1945 the United States was unquestionably the strongest power in the world. Its navy was more powerful than any two other existing fleets, Admiral William D. Leahy, then White House chief of staff, declared. America also enjoyed the best-equipped ground army, the greatest air force, and the secret of the atomic bomb, "the world's most fearsome weapon." Unlike all the other great powers fighting in World War II, the United States emerged relatively unscathed. Whereas Britain, China, France, Germany, Italy, Japan, and Russia had suffered devastating attacks on their homelands and lost millions of lives, the continental United States had been free of attack and relatively fewer—some 325,000—American lives were sacrificed in the conflict. Moreover, unlike its allies and enemies, the United States came out of the war with its productive facilities intact and with unprecedented economic might.

By dint of this new-found power, the United States seemed to be a law unto itself. There was little need to hear other voices or to follow any other nation's design. But most Americans still believed that this position was based less on might than on right. Wendell

Willkie's wartime best-seller, *One World*, reflected the mood. It described a shared desire around the globe for international cooperation on American terms—cooperation to promote democracy, free enterprise, and collective security through a new world league. In 1945 between 80 and 90 percent of Americans supported involvement in a world organization of this kind. According to this view, American power was not a source of intimidation but of inspiration; strength was an extension of the virtue of the American people and nation. It was assumed that other peoples around the globe shared this perception and were eager to follow the U.S. lead. "Americans tend to become political 'isolationists' when they cannot dominate international affairs and 'internationalists' politically when they can," the historian Walter LaFeber has written. His observation perfectly describes the state of affairs at the end of World War II. In a euphoric mood brought on by victory in the fighting and the prospect of unparalleled world dominance, Americans translated U.S. influence and control into universal agreement on how all nations should achieve international prosperity and peace.[1]

The feeling was bipartisan. The Roosevelt administration was joined by prominent Republican leaders like former President Herbert Hoover and Henry Luce, president of Time, Inc., in expressing these beliefs. They envisioned a peaceful, cooperative world in which allies and enemies alike would abandon old-style international politics—spheres of influence, alliance systems, balances of power—for participation in a new version of Woodrow Wilson's League of Nations. Colonial empires would disappear, and under the benevolent aegis of trustees appointed by the new world organization former colonies would move toward democratic self-determination. Most important, peoples and governments everywhere would become more like the United States—democratic, capitalist, and peace-loving.

The Soviet Union, China, Germany, and Japan all fit the bill. Communism was passing from the scene, Americans were urged to believe during the war. "Marxian thinking in Soviet Russia is out," the *New York Times* declared in 1944. "The capitalist system, better described as the competitive system, is back." Other commentators pointed to a steady movement away from a "narrow Marxian ideology in the direction of ideas that we can call, in very broad terms, democratic." Russia was no longer truly communist, Herbert Hoover told the Republican convention that year. (This optimism about the failure of communism has been a mainstay of American dealings with the Soviet Union for the past forty years. Although hopes of a Soviet transformation into a noncommunist society wax and wane, they never disappear. The West "will transcend communism," President Ronald Reagan declared in 1981. "We will not bother to denounce it, we'll dismiss it as a sad, bizarre chapter in human history whose last pages are even now being written.")

The Russians were also pictured during World War II as ready to jettison traditional power arrangements for a new structure of peace. The Yalta agreements, FDR said after signing them in a February 1945 meeting with Churchill and Stalin, "spell the end of the system of unilateral action, the exclusive alliances, the spheres of influence, the balances of power, and all the other expedients that have been tried for centuries—and have always failed."

China was similarly celebrated during the war as a "great democracy," a nation systematically turning into an East Asian United States. "China was not an alien country, full of strange customs, but a warm-hearted, hospitable land filled with friends of America," Willkie advised in *One World*. America's enemies, Germany and Japan, were also to be drawn into the democratic circle. After considerable argument among American officials, it was decided that rehabilitation rather than repression should be their postwar fate. A denazified, demilitarized Germany and a politically and economically reorganized Japan were to become New Deal–style democratic states.[2]

For most Americans these anticipated developments were the essence of international cooperation. A world becoming just like the United States was the prerequisite for an end to America's deeply ingrained habit of isolating itself from international politics. "I do not believe that any nation hereafter can immunize itself by its own exclusive action," the former isolationist Senator Arthur H. Vandenberg, Republican of Michigan, declared in 1945. "I want maximum American cooperation." But it was to be "cooperation" in a world where, as influential publisher Henry Luce put it, "American experience is the key to the future. . . . America must be the elder brother of the nations in the brotherhood of man."[3]

This pattern was first manifested in the economic agreements reached at Bretton Woods, New Hampshire, in 1944. The conference, attended by Britain, China, France, the Soviet Union, the

United Sttaes, and thirty-nine other nations, established the structure of institutions and rules for a stable and open system of international exchange and finance. Among the conference's institutional innovations was the International Bank for Reconstruction and Development (IBRD, commonly known as the World Bank), intended to promote postwar reconstruction in devastated industrialized countries and development in less economically advanced nations. Bretton Woods also established the International Monetary Fund (IMF), to ensure stable exchange rates between currencies and favorable conditions for world trade, which fluctuations in currency values could undermine.

Although ostensibly cooperative institutions aimed at serving the international good, the World Bank and the IMF had actually been designed in negotiations with the British before the conference, and they chiefly met American needs. British influence in these talks was minimal. According to Dean Acheson, then an assistant secretary of state, the British recognized that "we really are going to write the ticket, and all they wanted is . . . to be allowed to come in on the formulation from the start." Accepting the dominance of the dollar among world currencies, the transcendant role of the United States in international trade, and American control over the World Bank and the IMF, the conference reflected the realities of America's new economic power.

This, however, did not stop American leaders from trumpeting the results of the meeting as a significant gain for international understanding. "Commerce is the life blood of a free society," FDR had told the delegates to Bretton Woods. "We must see to it that the arteries which carry that blood stream are not clogged again, as they have been in the past, by artificial barriers created through senseless economic rivalries." If Congress did not approve the agreements, Acheson warned, "we might look with some apprehension upon the whole state of the world." In pressing the case for American participation, government leaders were frank to call attention to the benefits that would accrue to the United States—Secretary of the Treasury Henry Morgenthau, for example, predicted that the World Bank and the IMF would create an overseas market for one million American automobiles a year—but they sincerely believed, nonetheless, that these institutions would promote expanded production, employment, exchange, and consumption around the world.[4]

An emphasis on the international harmony and economic progress resulting from the conference was a concrete example of what American leaders had been promising for the postwar era, and it was calculated to encourage the triumph of internationalism over isolationism in the United States. But this was also a reflection of what most Americans hoped to sustain at home. Having achieved a substantial measure of national unity during the thirties and the forties, a majority of Americans, who remained more concerned with internal than external affairs, projected their wishes for the future at home onto conditions abroad.

"To the average American," the psychologist Jerome Bruner wrote in an analysis of wartime opinion, "the domestic and international are far from equivalent in either personal significance or interest. In spite of the years of war, the events and problems which beset the world beyond our boundaries are of secondary interest to the man in the street. To him the payoff is what happens right here at home—and what is likely to happen." Despite the fact that Americans had been involved in a global war, their attention largely remained fixed on what was happening inside the United States. This translated into a view of overseas affairs that reinforced a vision of a harmonious, cooperative America. Eager to maintain the sense of shared purpose that had blossomed during the New Deal and the war, most Americans boosted their faith in the durability of one America with illusions about the emergence of "one world."

By indulging themselves in this fantasy, Americans did more to undermine than to promote international accord. Describing the self-serving Bretton Woods agreements as a victory for internationalism, American leaders encouraged foreign convictions that the U.S. government could not distinguish between its own and international interests and that such unrealism—or even hypocrisy, as some believed it to be—made it particularly difficult for any other nation to deal on equal terms with the United States. To be sure, many foreign leaders recognized that American and international prosperity were inextricably tied together and that America's pursuit of its self-interest was no different from what they themselves did. Still, they were put off by American pretensions to altruism, especially when they heard conservative isolationists like Senator Robert Taft, Republican of Ohio, complain that joining the IMF would mean "pouring money down a sewer."

Seeing how beneficial the agreements were to American interests, foreign officials found it difficult to understand complaints that debtor nations would control the IMF and compel the United States to provide an endless supply of dollars while not giving up "one exchange restriction, one trade restriction, or one sterling area." Because America's wartime allies were mindful of how much the Bretton Woods commitments worked to America's advantage, they dragged their heels about implementing them, finally moving only under strong prodding from Washington. It was one thing to acknowledge and gracefully accept American economic dominance, but to paper it over with declarations of unlimited generosity was enough to make any thoughtful foreign observer a critic of the United States.[5]

In the context of America's extraordinary power and prestige in the closing months of the war and immediately after, however, it was impossible for most Americans to imagine any significant foreign grievance against the United States. Many assumed that nations everywhere, dazzled by American abundance, freedom, stability, and humane generosity, would be grateful for the U.S. contribution to the destruction of European and Asian tyrannies and would be inspired not only to imitate the American system, but also to cooperate in the construction of a new world order symbolized by the United Nations. In the 1980s, after forty years of superpower rivalry, regional wars, and unabating international strife, it is difficult to appreciate the intensity of wartime and immediate postwar hopes for global harmony and peace through a world league. But victory in the war and a taste of unprecedented world power for the United States had generated visionary dreams that now seem entirely out of reach.

In October 1943, when Secretary of State Cordell Hull attended a foreign ministers' conference in Moscow, he asserted the need for a four-power declaration on a postwar peacekeeping organization. "If any official in my country should announce that he were opposed to formulating the fundamental policies for a postwar world until after the war is over," Hull said, "he would be thrown out of power overnight." Although Hull persuaded the British and the Soviets to issue such a pronouncement, they were skeptical of plans for a new world organization. At the Tehran conference in November–December 1943, Stalin and Churchill raised questions about the American proposals, but Roosevelt disarmed their objections with warnings that the American public, and particularly Congress, would insist on a worldwide approach to keeping the peace. Similarly, at the Dumbarton Oaks talks between August and October 1944, when the Soviets demanded seats for all sixteen of their "republics" in the world organization and an all-inclusive veto for permanent members of an executive council, Roosevelt was shocked and warned that such proposals "would very definitely imperil the whole project."

Roosevelt's dismay was not so much at Soviet efforts to achieve national advantages; he himself expected the world body to be dominated by the "Four Policemen"—Britain, China, the United States, and the U.S.S.R.—and he had well-defined ideas about how the United States would protect its national interests after the war. But he understood that the great majority of Americans expected the realization of Wilsonian visions of universalism or collective security. Consequently, at Yalta he persuaded Churchill and Stalin to endorse his plans for a new international organization. In the last days of Roosevelt's life, when Stalin, in an effort to denigrate the importance of the nascent world body, indicated that Soviet Ambassador Andrei Gromyko, rather than Foreign Minister Vyacheslav Molotov, would attend the San Francisco organizing conference on the United Nations, FDR begged him to send Molotov. Since all the other sponsoring countries were sending their foreign ministers, Molotov's absence, he said, "will be construed all over the world as a lack of comparable interest in the great objectives of this conference on the part of the Soviet government." Only after Roosevelt died in April did Stalin agree, as an expression of respect for FDR, to send Molotov to San Francisco.[6]

Despite the realities of power politics and emerging Soviet-American differences over Poland and spheres of influence, which would shortly lead to the Cold War, Roosevelt had been determined to preserve the image of harmony that was so essential to internationalist commitments in the United States. Never mind how the world would be organized or how much postwar affairs would conform to traditional national priorities rather than universalist dreams, illusions about cooperative nations following America's idealistic lead needed to be preserved. For the Soviets, British, Chinese, and French, the American insistence on pursuing self-interest while describing the process as internationalism was a source of irritation that soured future dealings with the United States.

This is not to suggest that somehow the United States caused most of the postwar difficulties in the world, especially between itself and the Soviet Union. The Soviet Union's suspicions and its rigid determination to provide in every possible way for its own security, particularly in Eastern Europe; British hopes of clinging to the empire; Chinese efforts, by Nationalists and Communists alike, to use other nations to strengthen their own control over China; and French assertiveness in search of renewed national power—all were sources of international friction beyond U.S. control. Nevertheless, Washington's insistence on foreign conformity to ideas peculiar to American thinking and the U.S. effort to consign others to a junior partner's role in international relations stirred antagonisms.

Truman's dealings with the Soviets and the misunderstandings of 1945–46 that fed directly into the Cold War further illustrate the point. When Truman succeeded to the presidency in April 1945, Moscow and Washington were becoming increasingly divided over Poland. Roosevelt had had little expectation of winning genuine independence for the Poles from Soviet domination; but for the sake of American and world opinion, he pressed Stalin at Yalta into accepting a Declaration on Liberated Europe that created a different impression. In the weeks after Yalta, therefore, when the Soviets refused to allow a reorganization of the Polish government that would weaken communist control and lead to free elections, FDR became indignant, not because he expected Stalin to alter course and accept Polish self-determination, but because he believed that Stalin had committed himself to handling the Polish question with greater sensitivity to international opinion.

Truman, however, took the Yalta agreements at face value. Since he viewed them as fundamental commitments on which good relations between Moscow and Washington depended, he expected the Soviets to live up to the most important provisions. Moreover, for Truman, who warmly believed in Wilson's League and the necessity of a United Nations to head off future wars, it was essential that Moscow accept America's new world design, in which all nations would rely on collective international action rather than sphere-of-influence diplomacy to defend their security. Seeing this design not as an expression of U.S. dominance over international affairs but as the genuine triumph of a new and far better world system, he was outraged by Soviet disregard for American wishes on Poland.

In an initial meeting with Molotov in late April, Truman described the Polish question as a major symbolic issue for the American people. When it became clear that his appeal had had no impact on the foreign minister, Truman complained privately that agreements with the Soviets were a one-way street, and that if they did not wish to join the United Nations, they could "go to hell." Reinforced in these feelings by advisers like Averell Harriman, then ambassador to Moscow, who pictured Soviet actions as a "barbarian invasion of Europe," and Secretary of the Navy James Forrestal, who urged a showdown with Russia sooner rather than later, Truman decided to "lay it on the line" with Molotov. Telling the foreign minister that the United States would not be a party to the formation of a Polish government that was undemocratic, the president warned that postwar economic assistance to the Soviet Union could not be divorced from the Polish issue. When Molotov refused to budge, Truman retorted that an agreement had been reached on Poland and it only remained for Stalin to keep his word. Molotov's complaint that he had "never been talked to like that in my life" provoked Truman to reply, "Carry out your agreements and you won't get talked to like that." "I gave him the one-two, right to the jaw," the president later said.

It would be a mistake to exaggerate the importance of Truman's bluntness. It is doubtful that he could have said anything that would have altered significantly the course of Soviet actions in Poland or prevented the further deterioration of Soviet-American relations. Even if the United States had been prepared to concede openly a Soviet sphere in Eastern Europe and to share the secret of the atomic bomb with Moscow, it is unlikely that two countries with such diametrically opposed economic, political, and social systems would have remained on good terms once the wartime alliance had run its course. Still, this should not obscure the fact that Truman's actions were the product of some serious misunderstanding on his part. He was insensitive to the importance of Poland in the Soviet view of its own security; it was hard for Stalin to imagine that the United States had no ulterior motive for interfering in Polish affairs. Blinded by the same national self-righteousness that moved Americans to expect adherence by Europe to the U.S. view of world affairs, Truman heightened the antagonism which was beginning to dominate Soviet-American relations.[7]

A similar point can be made about the administration's handling of relations with China. At the close of World War II the renewal of Nationalist-Communist hostilities diluted American hopes for a unified, democratic, internationally cooperative China. Convinced that the United States could make a decisive difference in China's civil strife, Truman sent General George C. Marshall to Chungking to mediate differences between the two sides. When Marshall won an initial agreement to a truce, Truman expressed hopes that "we will come out with a unified China and a good friend in the Far East for the United States." But as Truman, Marshall, and others in the administration came to realize during the next year, it was impossible for the United States to influence the outcome of China's great internal struggle. That was a conclusion, however, which many politically active and influential Americans refused to accept, and so the administration continued to give Chiang Kai-shek's Nationalists economic and military aid well after it was clear that their cause was lost. The consequence was greater estrangement from the emerging Chinese communist government than there need have been. Less concerned with Chinese communist views of American policy than with politically explosive feelings in the United States about preserving a "free" China, Truman accommodated the parochial vision of an America capable of bending China to its will. A principal cost of this shortsighted policy was the legacy of hostility between Peking and Washington that endured for more than twenty years.[8]

Similar difficulties surrounded the Truman administration's policy on the international control of atomic power. At the start of his term Truman decided to follow Roosevelt's decision to deny the Soviets information on the bomb. At the same time he also saw the need for "international arrangements" to prevent the postwar development and use of more atomic weapons. In November 1945, therefore, Truman joined with the British and Canadian prime ministers, Clement Attlee and Mackenzie King, in proposing that international control of atomic energy be turned over to a U.N. commission. The objective was to encourage nations to accept successive regulatory measures: the exchange of scientific information, an agreement on peaceful uses of atomic energy, the elimination of atomic weapons, and international inspections to prevent national violations of agreements. During the many years it would take to complete this process, the United States was to maintain its monopoly of existing atomic bombs. If this seemed like a generous proposal to Truman, who had to worry about conservative U.S. senators objecting to any scientific exchange as "sheer appeasement" of Russia, the Soviets wanted no part of a program that would even temporarily sustain America's unilateral control over atomic weapons.

In the atmosphere of worsening relations with the Soviet Union, which maintained a significantly larger army than did the United States, it was politically impossible for the Truman administration to give up its atomic advantage. Consequently, when the United Nations established the Commission on Atomic Energy, which was to convene in June 1946, the administration sought ways to advance international cooperation without giving up America's exclusive hold on the bomb. It was like trying to square a circle. An initial "liberal" proposal by then Under Secretary of State Dean Acheson and David Lilienthal, chairman of the Tennessee Valley Authority, urged creation of an international Atomic Development Authority which would control all raw materials used in the development of atomic power, including those in the Soviet Union, and would bar all nations from making new atomic bombs, while those already in American hands would be maintained. To make such an arrangement more palatable to the Soviets by demonstrating trust in them, the proposal minimized international inspections and said nothing about punishing treaty violations.

At best, this plan had little prospect of being accepted by Moscow, which was less than four years away from developing its own bomb and unwilling to submit to outside control of any Soviet resources; but its rejection was made certain by the appointment of Bernard Baruch as the U.S. representative to the U.N.'s Commission on Atomic Energy. A conservative Democratic financier with a reputation as a staunch anti-Communist, Baruch was described by Lilienthal as someone whom the Russians would suspect of trying "to put them in a hole, not really caring about international cooperation." Indeed, Baruch immediately amended the Acheson-Lilienthal proposal to include sanctions against violators of the agreement and to suspend a nation's right to veto any punishment voted against a treaty violator by the U.N. Security Council.

Baruch "had hardened U.S. opinion behind ideas that made no sense," journalist Walter Lippmann complained at the time, and unnecessarily antagonized the Russians. "The creators of the Baruch plan," historian Gregg Herken later concluded, "guaranteed that international control would be entirely on American terms—or not at all." Yet in spite, or perhaps because, of these conditions, reaction in the United States to the Baruch plan was generally favorable. Depicted as a proposal for cautiously sharing American secrets about atomic energy and ultimately giving up atomic bombs, it was celebrated as a demonstration of American generosity and cooperativeness in trying to avert a future nuclear war. Most newspapers emphasized American readiness to make sacrifices for peace. Conservative opponents of the plan did not dispute this interpretation, but complained that it risked giving up American secrets and power to "foreign masters."

Other governments, however, saw Baruch's proposal as a thinly disguised attempt to ensure America's continued monopoly of what he called "the winning weapon." They saw confirmation of this design in America's rejection of a Soviet call for international prohibition of the production, possession, or use of atomic bombs, and in America's first postwar atomic test on July 1, 1946, only seventeen days after Baruch presented his plan. In the view of one member of Baruch's staff, who resigned in protest against American atomic policies, "other nations do not regard our bomb monopoly as the exercise of a sacred trust but as an imminent threat to their existence." What appeared to most Americans as an act of international goodwill impressed many foreigners as a dangerous reach for national power that could lead to an arms race and a possible nuclear war.[9]

The veneer of internationalism covering America's largely unilateral control in the Bretton Woods agreements and the Baruch plan was also extended, with similar results, to American leadership of its allies in other conflicts with the Soviets. Although the European opponents of communism and expanded Soviet power greatly valued America's defense of their independence, they distrusted the American claim to be constructing a cooperative enterprise in which equal partners were combating a common menace. In March 1946, for example, when Winston Churchill described the descent of an "iron curtain" across Europe, he also urged "a fraternal association of the English-speaking peoples" to meet the Soviet challenge. By suggesting an Anglo-American alliance outside of the United Nations, Churchill provoked considerable antagonism from American internationalists like Eleanor Roosevelt and liberal senators Claude Pepper, Harley M. Kilgore, and Glen Taylor. Ten days after Churchill's speech, when he appeared in

New York, pickets marched outside of his hotel chanting, "Winnie, Winnie go away. U.N.O. is here to stay," or "Don't be a ninny for imperialist Winnie."

Opposition to Churchill's plea for an old-fashioned alliance against the Soviet Union and to the new "get tough with Russia" attitude in the United States was voiced most forcefully by Secretary of Commerce Henry Wallace, the former vice-president who between 1945 and 1948 became the chief spokesman for liberal Democrats trying to preserve FDR's vision of postwar harmony. In a speech in New York in September 1946, Wallace made the case for a restoration of better relations with the U.S.S.R. and a return to the cooperative spirit fueling the drive to establish the United Nations.

For his troubles, Wallace was dismissed by Truman, who by then saw no possibility of working closely with the Soviets. Nor, however, was the president prepared to follow Churchill's lead. The Truman administration stood between Churchill, who was calling for Anglo-American cooperation against the Soviets, and America's liberal internationalists, who were urging the continuation of cooperation with Russia through the United Nations. Although Truman was now intent on meeting the Soviet challenge around the globe, he preferred to issue unilateral declarations, to which like-minded allies were supposed to adhere spontaneously as members of what Americans now began to call the "free world."[10]

On March 12, 1947, Truman signaled the shift from America's wartime universalism, in which all nations were supposed to work with the United States through the United Nations, to a more limited international cooperation of anti-communist states. In a renowned speech that came to be known as the Truman Doctrine, the president called for a worldwide crusade guided by the United States:

> At the present moment in world history nearly every nation must choose between alternative ways of life. . . . One way of life is based upon the will of the majority, and is distinguished by free institutions. . . . The second way of life is based upon the will of a minority forcibly imposed upon the majority. It relies upon terror and . . . the suppression of personal freedoms. I believe that it must be the policy of the United States to support free people who are resisting attempted subjugation by armed minorities or by outside pressures.

The immediate motive for Truman's address was a crisis in the eastern Mediterranean. The British had been supporting a monarchical government in Greece against a communist-led rebellion, as well as a conservative regime in Turkey that felt threatened by the Soviet Union. When the British government informed Washington that its own economic difficulties would not permit it to supply Greece and Turkey with sufficient economic and military aid to withstand the pressures, the Truman administration felt compelled to come to the rescue. A central purpose of Truman's speech, therefore, was a request to Congress for $400 million to help keep Greece and Turkey out of the Soviet orbit. As Acheson privately told congressional leaders, the stakes were not simply Greek and Turkish independence but the freedom of the Middle East, South Asia, Africa, and Europe. This was ultimately a battle between liberty and totalitarianism, Acheson said, in which the United States was defending itself against a worldwide Soviet Communist drive for power and control.

The challenge to Western interests at that moment was real, and the United States needed to respond, which it did with loans and military supplies that helped the anti-communist governments in Greece and Turkey survive. But as some journalists and policymakers at the time and numerous historians since have pointed out, there is reason to question the all-inclusive, evangelical quality of the Truman Doctrine. A "global policy" that aimed at aiding "free peoples everywhere" impressed Walter Lippmann as excessive. A prominent columnist who had helped shape the Fourteen Points, Woodrow Wilson's program for world peace, and who now urged the need for "realism" in foreign affairs, Lippmann thought it "better to address the Soviet Union directly" and adopt "a precise Middle Eastern policy. . . . A vague global policy, which sounds like the tocsin of an ideological crusade, has no limits. It cannot be controlled. Its effects cannot be predicted. Everyone everywhere will read into it his own fears and hopes. . . ."

Political scientist Adam Ulam observed years later that

> it would have been vastly preferable had Truman been able to present the grim realities of the postwar world more frankly: to say that it was in the interest of the United States that Greece and Turkey not be Communist powers; and that while the Greek and Turkish governments did not conform to the concept of democracy as understood in Britain and the United States, they still were preferable for their own people and for the world to Communism as practiced by Stalin. . . .

Such candor, Ulam said, might have avoided "much of the ensuing confusion . . . as to which nation the United States should help and why." It might have put a damper on the impulse to assert American power against Communists everywhere, and it might have eased rather than inflamed suspicions abroad that the United States would oppose revolutions in the name of U.S. security interests regardless of their merits. But Truman could not escape history. His doctrine was essentially an extension of the crusading fervor that had informed every major American foreign policy initiative in the previous fifty years. To win wide public support, a call to action abroad had to be described as a contest between good and bad, a struggle to save the world from evil.

Consequently, the Truman Doctrine intensified fears that America lacked the ability to discriminate between nationalist and communist opposition to the status quo in Asia, Africa, Latin America, and the Middle East, or to understand that a left-wing regime in an underdeveloped country did not automatically imply affiliation with Moscow. Revolutionaries in Algeria, Guatemala, and Vietnam, for example, could hardly have had much confidence that Truman's speech meant the continuation of America as "the victorious emblem around which may rally the multitudes thirsting for social justice." On the contrary, it cast the United States in the role of a counterrevolutionary power preserving conservative regimes. The contribution of the American government to the preservation of the conservative Greek and Turkish regimes in the late 1940s strengthened this view of the United States. A less doctrinaire address might not have mobilized the widespread domestic political support Truman and others thought essential to the administration's immediate goal, but it would have made foreign policy errors and antagonism to the United States less likely in the long run.

There were also some short-term negative repercussions of Truman's overdrawn rhetoric for U.S.-Soviet relations. In part, the American expression of determination to hold its ground was highly salutary: At the Big Four foreign ministers' conference in Moscow in April 1947, the Soviets seemed more flexible about a German peace settlement, proposing the creation of a unified, demilitarized, neutral German state. At the same time, however, they demanded $10 billion in reparations and a share in the supervision of Germany's industrial Ruhr Valley. They also filled the air with accusations against the United States and hinted at, but did not explicitly ask for, a summit conference of heads of government. One can only speculate about how much more receptive the Soviets might have been to a German agreement had the Truman administration been less bombastic in its declaration of the American intention to stand fast in Greece and Turkey. Had Truman addressed questions of American and Western interests in a more businesslike manner, it is conceivable that Moscow might have responded in kind. As it was, though, when Truman followed Senator Vandenberg's advice to "scare the hell out of the country" as a way to convince people "that the war isn't over by any means," the president also must have frightened the Russians into more determined efforts to oppose American power and influence everywhere.[11]

Yet opportunities for reducing Soviet-American tensions continued to exist. When Secretary of State George Marshall discussed a German settlement with Stalin in Moscow in April 1947, the Soviet leader expressed the belief that compromise was still possible. Marshall, however, saw Stalin's comments not as an opening to further talks but as a sign of Soviet determination to prolong discussions while Europe moved toward economic collapse. Having just sat through weeks of discussion with Molotov, in which the latter showed himself to be unwilling to reach a compromise settlement on European problems, Marshall had little faith in Soviet interest in an agreement with the West. When he returned to the United States, Marshall warned publicly that Europe was "disintegrating" rather than recovering from the war: "The patient is sinking while the doctors deliberate. . . . Whatever action is possible . . . must be taken without delay."

Marshall's concern gave rise to the plan bearing his name. In June 1947 he proposed an economic recovery program designed by the Europeans themselves and agreed to "by a number [of], if not all European nations." He and others in the administration officially urged Soviet and East European participation, partly as donors but principally as recipients of aid. Yet they set conditions for involvement that made it highly unlikely that Moscow and its satellites would take part. Insisting on international planning and an exchange of information that would have subjected the Soviet economy to Western scrutiny, the Truman administration guaranteed that Moscow would reject Marshall's overture. Because Russian participation would have greatly increased the cost of the program and aroused conservative opposition in the Congress, where the Republicans were now in control, the administration was just as happy to foreclose Soviet involvement from the start. But unwilling to take the blame for having divided Europe by openly excluding

Moscow, the administration paid lip service to the idea that no nation should be denied a chance to participate.

To everyone's surprise, however, the Soviets initially showed some interest in taking part. In response to an Anglo-French proposal for a three-power meeting on Marshall's plan, Moscow sent a delegation headed by Molotov to Paris at the end of June. The scheme held enticements that Stalin found hard to resist. According to Soviet expert Adam Ulam, the Soviet leader may have seen Marshall's proposal as a chance to speed the repair of Russia's drastically damaged economy with U.S. aid and to accelerate "that industrial advance which was so high on the list of Stalin's priorities." Yet the Marshall Plan "was a gamble which Stalin could not take," says Ulam. It would have meant opening up the Soviet Union to observation of its economic condition, and the Soviet leadership worried that this might embolden Moscow's enemies to challenge its grip on Eastern Europe. The Russians broke off negotiations with the British and French in the first week in July, announcing, candidly enough, that they wanted no part of a program which would infringe on national sovereignty. The real goal of the Marshall Plan, they soon declared, was not economic recovery or a defense of the West against communism but the destruction of socialist regimes in Eastern Europe. One early, unintended effect of the Marshall Plan, then, was the deepening of Cold War tensions between East and West.[12]

The principal thrust of the plan, however, was a necessary and bold initiative for overcoming economic problems in Western Europe. Despite Soviet suspicions, it was essentially a defensive program for strengthening Western economies against potential Communist subversion encouraged by the U.S.S.R. and a means of securing European markets for American goods.

Yet because the Marshall Plan was more a defensive response to possible indirect Soviet aggression than an assault on Moscow's power or a disguised effort to induce dependency on the United States and some loss of European sovereignty, it might have done less to intensify Soviet-American difficulties if Washington had been as forthcoming with Moscow as its rhetoric suggested. If the sensitivity demonstrated by Acheson and Lilienthal to Soviet secretiveness in outlining plans for international control of atomic energy had been duplicated in the Marshall Plan, that might have persuaded the Soviets to participate in the economic reconstruction of all of Europe. Obviously, such flexibility and generosity on the part of the Truman administration would have provoked an outcry among conservative Americans that the United States was giving aid to a sworn adversary. Undoubtedly, it also would have provoked references to Munich and the lessons to be learned from the appeasement of Hitler in the thirties. But the Munich analogy was not particularly useful for dealing with the Soviets in the immediate postwar years. Wise statesmanship would have made allowance for the fact that an exhausted Soviet Union was not comparable to an aggressive Nazi Germany. The long-term gains involved in disarming Soviet suspicions and drawing them into a dialogue with the West might have made it worth taking the limited risks of a more flexible economic proposal.

Walter Lippmann emphasized the last point at the time. Though a European recovery formula "could be made to work without eastern Europe and without Soviet collaboration," he wrote in June 1947, it "would work far better if eastern Europe participated and the Soviet Union approved." The fundamental idea was "not to form a Western bloc which excluded eastern Europe and was defensively and offensively opposed to the Soviets —but to form the nucleus of a European union which is meant to include them, and offers them superior advantages if they collaborate than if they do not."

Lippmann's criticism of the Marshall Plan was a prelude to a wider debate he subsequently joined with the Truman administration generally and George F. Kennan, the head of the State Department's Policy Planning Staff, in particular, concerning American policy toward the Soviet Union. In July 1947 Kennan, under the pseudonym "X," published a landmark article in the quarterly journal *Foreign Affairs* entitled "The Sources of Soviet Conduct." Describing Soviet behavior not as the product of rational security needs shaped by Russian history but as the expression of a messianic Marxist-Leninist ideology and a paranoid sense of insecurity, Kennan urged that the Soviets be "contained by the adroit and vigilant application of counterforce at a series of constantly shifting geographical and political points." In sum, the West faced an inexorable drive for world power by a formidable foe that could only be countered by "unanswerable force."

This, said Lippmann, was a misreading of Russian intentions and was certain to freeze the two sides into a prolonged and unnecessary struggle. Describing Soviet conduct as more the consequence of Russian history than Marxist ideology, Lippmann asserted that the

Russians were in Eastern Europe because, having suffered repeated invasions from the west which cost them dearly, they were determined to close off the corridor for a fresh attack. The way to deal with Moscow, he said, was not through containment "at a series of constantly shifting" peripheral points but by reaching a political settlement in Europe. He believed that if the Soviets hoped to subvert Western European regimes and replace them with sympathetic communist governments, it was largely because they felt threatened by U.S. actions in Europe. An offer from the United States for mutual withdrawal of American and Soviet troops from central Europe and for unification of a neutral, demilitarized Germany seemed to Lippmann likely to narrow the widening breach between the United States and the U.S.S.R. and to advance the cause of long-term peace.

For Lippmann, the appropriate goal was not to flex military muscles and meet a nonexistent threat, thus increasing the likelihood of war, but rather to reduce Soviet-American antagonism through diplomacy. "The history of diplomacy," he observed,

> is the history of relations among rival powers, which did not enjoy political intimacy, and did not respond to appeals to common purposes. Nevertheless, there have been settlements. Some of them did not last very long. Some of them did. For a diplomat to think that rival and unfriendly powers cannot be brought to a settlement is to forget what diplomacy is all about.

Lippmann's was a dissenting voice in 1947, as Kennan's has been since the 1950s, when he began advocating a number of Lippmann's ideas.[13]

Lippmann understood how the United States unwittingly contributed to its own alienation from the outside world. American errors in foreign policy during and immediately after World War II were not the product of malicious intent or conspiratorial schemes to impose control on the world. Nor does the United States suffer from such a collapse of America's traditional anti-imperial standard today, as critics of recent U.S. policies contend. Rather, the difficulty was and remains a less obvious one: a failure of imagination and courage to think in bolder, less categorical or conventional ways about international affairs.

During the war Americans deluded themselves into believing that all nations wished to follow their example, that they had an irresistible design for the future. It was assumed that the American system of beliefs, more than U.S. power, would attract other governments to cooperate as junior partners in a new world league. Although difficulties, particularly with the Soviets, signaled the limitations of such a world view, the Truman administration reiterated the goal in the Baruch Plan, a scheme for sharing atomic power under America's benevolent guidance. When that proposal only served to exacerbate the problem, the American government responded to a crisis in the Middle East with a declaration of ideological warfare. Although the Marshall Plan initially included a bow in the direction of the earlier universalist design, it became a prototypical Cold War program for defending the West against significant Soviet advances in Europe and, as such, helped intensify the Cold War.

Had the Truman administration been less doctrinaire about the Soviet or, more broadly, the communist threat, there might have been a narrower gap between the United States and the U.S.S.R. after 1947, and between the United States and other national states that came into existence after the war. To be sure, there would have been no utopian settlement. Abundant problems, requiring sustained thought, attention to detail, and painful effort, would still have dogged U.S. relations with Moscow and the rest of the world; but there might have been some significant gains, such as reductions in colossal defense costs and global tensions. These would have meant taking political risks at home and abroad; domestic opinion was unreceptive to bold initiatives in foreign relations, and the Soviets tended to read sincere overtures as signs of weakness or deviousness. Nevertheless, the stakes—international agreement and peace—were sufficiently high for the United States to have been justified in taking these risks and, in so doing, declare itself ready for a more subtle and effective response to events around the world in the years ahead.

3 The Korean Vortex

ROBERT J. DONOVAN

Robert J. Donovan, formerly Washington Bureau chief of the *New York Herald Tribune* and later of the *Los Angeles Times*, is the author of two volumes on the Truman presidency and other books, including *Nemesis: Truman and Johnson in the Coils of War in Asia*. He was a fellow at the Woodrow Wilson International Center for Scholars in Washington in 1978–79, senior fellow at the Woodrow Wilson School of Public and International Affairs at Princeton University in 1979–80, and Ferris Professor of Journalism at Princeton in 1980–81.

IN SEPTEMBER 1945, the turbulent month of the Japanese surrender aboard the U.S.S. *Missouri* in Tokyo Bay, an event much less dramatic took place a few hundred miles away, on the other side of the Sea of Japan: The United States XXIV Corps arrived off the southwestern coast of Korea and prepared to occupy that country south of the thirty-eighth parallel.

Until that moment, the American people historically had paid little attention to the remote land of Korea, unhappily wedged among the great powers of China, the Soviet Union, and Japan. But now American troops were going ashore, even as Soviet troops earlier had begun crossing from Manchuria into Korea to occupy the area north of the thirty-eighth parallel. By mid-1945, occupation of war-ravaged countries by the victorious powers in World War II had become commonplace. The idea was that the occupiers would stay for a limited time, administer, rehabilitate, and then depart, leaving behind a stable new order. That Germany would remain divided was unthinkable. And few would have guessed in mid-1945 that the double occupation of Korea would amount to much more than that. In fact, it was destined to have calamitous consequences.

Almost from the beginning of the occupation, Korea was turned into a cockpit of the Cold War, the embodiment of all the painful phenomena that skewed international relations after World War II. The list was a long one: a potentially endless Soviet-American arms race; pervasive intransigence and xenophobia on the Soviet side as against containment and fierce anti-communism on the American side; Soviet expansionism and meddling as against American acquisition of bases, expansion of air power, and stockpiling of nuclear arms; Soviet exploitation of revolutionary impulses in the developing world versus the American quest for economic gains and alliances aimed at communist powers; and, on both sides, ambition, delusion, miscalculation, suspicion, and mistrust.

For the current generation of Americans, the agony of the more recent Vietnam War, with its devastating impact on society, tends to blot out memories of the Korean War of 1950–53. Yet the Korean conflict, then the third-largest foreign military engagement in American history, was one of the worst anywhere in the twentieth century. It was a savage, costly, and senseless struggle, alike in its disruption at the time and in the ominous legacies it left. In addition to all the Korean lives that were lost on both sides, 32,629 Americans were killed in action in Korea, 20,617 died of other causes, and 103,284 were wounded.[1]

Already in a worrisome state by the end of World War II, Soviet-American tension continued during the Korean conflict to escalate toward dangerous heights, from which little permanent relief has ever been found. The war was the first in a chain of events after 1945 that awakened Americans (and the rest of the world) to the practical limits of U.S. conventional military power, which many had come to regard as virtually unchallengeable during the deployment of massive land, sea, and air forces in World War II.

It was China's intervention on the side of North Korea that implanted the sense of limitation. Fear of a similar Chinese intervention in Vietnam in the 1960s, for example, led President Lyndon B. Johnson to rule out an American invasion of North Vietnam and to restrict the American bombing of North Vietnamese targets. The American failure in the Vietnam War, in turn, produced the so-called Vietnam syndrome that in recent years has dampened the enthusiasm of Americans, notably including the military, for intervention in foreign crises.

Korea was a particularly dangerous conflict because, at the rawest moment of the Cold War, while Joseph Stalin still ruled the Kremlin, American fighters and bombers, American combat ships, and the ground forces of eight U.S. Army and Marine divisions were in action very close to the borders of the Soviet Union and Communist China. There were almost limitless possibilities for grave incidents, especially along the Yalu River, which formed the boundary between Chinese-controlled Manchuria and North Korea.

At the depth of the war, when the Chinese army intervened on the side of the North Koreans, General of the Army Douglas MacArthur, the U.N. commander, wanted to bomb China's industrial centers and blockade the Chinese coast, an operation which would necessarily have included a blockade of the Soviets' main Far Eastern port, Vladivostok. Such a blockade would have been an act of war against China and the Soviet Union.

The conventional bombing of Chinese cities and a coastal blockade would not have been the extent of the punishment the United States would have inflicted on China, if U.N. troops had been forced to evacuate Korea, as it was feared they might be. If, in that circumstance, China had tried to press its advantage by bombing military ports and airfields in Japan, where the evacuating troops would have taken shelter, the strong possibility existed that President Harry S Truman would have retaliated with a nuclear attack on China. And, given the mutual defense treaty between China and the Soviet Union, that could have embroiled the United States in a full-scale war with both the Chinese and the Soviets.

Much about the origins of the Korean War remains obscure. The diplomatic records of the Soviet Union, China, and North Korea have never been published. Journalist I. F. Stone argued years ago for holding open the question of whether South Korea may have provoked the war, perhaps with the connivance of certain American officials, such as MacArthur or John Foster Dulles, but no proof of this view has ever come to light. Scholarly work today identifies North Korea as the perpetrator of the war, a conclusion furthered by the disclosure in Nikita S. Khrushchev's memoirs that the Soviets and Chinese assented to North Korean wishes to invade the South.

North Korea's invasion in June 1950 was the first time that a communist country aligned with the Soviet Union launched a military attack across a recognized international boundary. The response by the United States was the first American effort to contain communism with armed force. The policy of containment by economic and political means had already been instituted in Europe with the Truman Doctrine in 1947, and it had clearly been enunciated in the Far East when the Communists completed their conquest of China in 1949. At that point Secretary of State Dean Acheson ordered a new study of Far Eastern policy under the direction of Ambassador-at-Large Philip C. Jessup, instructing him to "take as your assumption that it is a fundamental decision of American policy that the United States does not intend to permit further extension of Communist domination on the continent of Asia or in the southeast Asia area."

Oddly enough, momentous as it was, the Korean War touched the lives of most Americans lightly. Selective service was extended, draft calls enlarged, and some reserves called up, but no general mobilization took place. Truman imposed wage-and-price and other economic controls, and he pushed three bills through Congress to increase taxes. But these measures had a stabilizing effect. After an initial burst in the summer of 1950, inflation was kept remarkably in check, and without acute consumer shortages the war years were generally a prosperous time at home. Since the Korean conflict began less than five years after the end of World War II, average Americans did not have a hard time adjusting to the idea of being at war, as they did later, at the time of Vietnam.

An overwhelming majority of Americans supported Truman's decision to commit armed forces to Korea. The reason was a popular belief—fostered by the government, both major political parties, and the press—that a firm line had to be drawn against further Communist expansion in areas then considered vital to the security interests of the United States. As would later be the case in Vietnam, public support for involvement in Korea persisted until the casualties and other costs multiplied while no sign of early victory appeared. Then the pendulum of public opinion swung heavily the other way. Nothing exasperates Americans like stalemate.

Even after the exasperation had taken hold, there was scarcely any public uproar or dramatic protest. For one thing, the Korean War came four years before the pivotal Supreme Court decision in *Brown* v. *Board of Education* outlawing racial segregation in the public schools and validating the struggle for civil rights. Not only was the United States still largely segregated in the early 1950s, but also the methods of popular dissent—marches, sit-ins, teach-ins, protest songs, and other challenges to authority—were still relatively unfamiliar at the time of the Korean War. These were still the quiet fifties, when the battles of Birmingham and Selma, not to mention Chicago and Kent State, were unimaginable. So during the Korean War no crowds in blue jeans surrounded the White House or besieged the Pentagon, as they did to protest the war in Vietnam. No one chanted at Truman an earlier version of "Hey, hey, LBJ, how many kids did you kill today?" There was no denunciation of "Amerika" as an heir to Nazi Germany. No students waved the North Korean flag in the streets or burned the Stars and Stripes along with their draft cards. Patriotism was not scorned, nor were American institutions assailed. This was true in part because the war remained remote. Battles and corpses did not appear in living color on home television screens. Those few families who did watch television saw only bloodless black-and-white newsreel shots on fifteen-minute-long evening news programs.

What set the stage for the general American acceptance of military challenge in Korea was a fear and detestation of communism, long in the making but especially intense by the middle of the twentieth century. True, the United States had been born of revolution. Living under a liberal constitution, however, and cherishing freedom of speech, religion, and enterprise in a bountiful land, the American people had grown hostile to anything that represented a challenge to established order, especially any collectivist philosophy that seemed to question, and might ultimately threaten, their own cherished individualism and liberty. After World War I, American animosity focused on Lenin, the Bolsheviks, and the growth of Soviet communism. The notorious domestic "Red Scare" of 1919 and 1920, while it angered many people, reflected strong currents in public opinion. During the Great Depression many intellectuals and some workers saw communism as an attractive alternative, but never did it become even a remote menace to security at home. During the war against Hitler, American anticommunism was subdued because of the necessity of an alliance with the Soviets. But the virus lived, and the Cold War caused it to flare again.

Long before Truman came to power, American mistrust of Soviet aims had begun to shape the ill-fated future of Korea. Indeed, Korea was a striking example of how the upheavals of world wars create new situations that lead to lesser wars later on. In the reshuffle of power caused by World War II, the Japanese empire toppled while the two new superpowers, the United States and the Soviet Union, emerged. As Japan collapsed, Korea, which had fallen under Japanese control in 1905 and was formally annexed in 1910, was set free, albeit in chaos. The two superpowers, with rival ideologies and goals, were contending for influence in the Far East. Unexpectedly, Korea became the locus of a conflict between them, carried on by their armed client states in the North and South.

During World War II a considerable number of Japanese troops had been stationed in Korea. For the so-called Grand Alliance, therefore, the status of postwar Korea was an inevitable consideration. At the Cairo Conference of 1943, Roosevelt, Churchill, and Generalissimo Chiang Kai-shek of China stated as a matter of policy—later adhered to by Moscow—that "in due course Korea shall become free and independent." But how was that declaration to be translated into policy by Roosevelt's State Department? In his outstanding work on the origins of the Korean War, Bruce Cumings recalls that two separate American views came into play:

> One vision placed emphasis on the capitalist virtues of free trade, open systems, the workings of the world market, and on the progressive virtues of representative democracy, aid to the downtrodden, and a generous if paternal sharing of American blessings. The other emphasized a more rough-hewn American impulse toward asserting national interest, carving territorial spheres and national economies, and confronting enemies of the American way.[2]

Roosevelt favored a multilateral trusteeship to guide Korea out of its colonized and exploited status. Such a trusteeship would be intended, according to Cumings, "to provide a benevolent condominium that would succor postcolonial peoples toward independence while maintaining an American foot in the door." But the American foot would not be the only one in the door; with its resurgent interest in the Far East, the Soviet Union also would be a trustee, along with China and Great Britain.

As early as 1943, however, the prospect of a Soviet presence in Korea began to arouse distrust in the State Department. Analysts including John Carter Vincent, William R. Langdon, H. Merrill Benninghoff, Hugh Borton, and Alger Hiss wrote policy papers reflecting their concern about future Soviet domination of Korea. (Ironically, Vincent and Hiss were later accused of being dangerously pro-communist.) Various undesirable possibilities suggested themselves to the policymakers: the ultimate Sovietization of a colonized people, the Soviet acquisition of ice-free ports and access to Korean resources, and a Soviet hold on a seemingly strategic area in the North Pacific. Already 35,000 Koreans had fallen under Soviet influence, after they fled north into Manchuria to escape Japanese rule, and some American officials felt that their return, after being indoctrinated into communism, itself represented a threat. The State Department viewed total Soviet domination of Korea as a threat to American security interests in the Far East and recommended United States participation in the occupation of Korea. The second half of the twentieth century was to bear the burden of Washington's ultimate acceptance of the thesis that first Korea and then Vietnam were vital to American security.

The situation in Korea was a microcosm of the disarray of much of the world in the years after World War II. That conflagration had simply obliterated much of the old order, throwing entire regions into turmoil. Germany and Japan were in ruins. Great Britain, though a victor, was hopelessly overextended, its resources nearly exhausted. The British Empire was slipping away in India and the Middle East. The French, the Italians, and other West Europeans were unable to restore their economies from the war's devastation. In stages, Eastern Europe came under Soviet domination. The Middle East was changed beyond recognition by the exploding struggle of the Zionists for a Jewish homeland in Palestine. China was in the grip of one of the greatest revolutions in history, and the sparks of nationalism also ignited disorder in Vietnam, the Philippines, Indonesia, and Malaya; in many Western eyes, these Asian upheavals were viewed to a large extent in terms of a struggle between communism and freedom.

At the time of his death on April 12, 1945, Roosevelt's postwar policies were in flux. When Truman was suddenly pitched into the White House at the climax of World War II, scant time was available to him for contemplation of secondary problems like Korea. Eventually, he accepted the idea of trusteeship. But, unlike Roosevelt, who conducted diplomacy to his own tune and usually ignored the professionals, Truman, unfamiliar with the handling of foreign affairs, was forced to rely heavily on the State Department, where the idea of the strategic importance of Korea to the United States had incubated. During the rush of events brought about by the atomic bombs and the sudden Japanese surrender, Truman and James F. Byrnes, his first secretary of state, were eager to consolidate American gains in the Far East, and they wanted to influence as large a part of Korea as possible. Hastily, Dean Rusk, the future secretary of state, who was then a colonel on the War Department general staff, and his Pentagon colleague Colonel Charles H. Bonesteel III drafted a proposal. They recommended that the dividing line be drawn at the thirty-eighth parallel. To Rusk's surprise, the Soviets accepted that recommendation. Truman decided to send occupation troops into southern Korea, primarily to contain the Soviets along a supposedly temporary boundary. (The boundary, of course, became permanent, and forty thousand American troops are still stationed below the thirty-eighth parallel with a mission that has not changed much in forty years.)

When Roosevelt had proposed trusteeship, he had in mind the

American experience in the Philippines, which the United States had conquered and then, over half a century, "prepared" for independence. But Roosevelt and his successors miscalculated the desire and even the tolerance of colonial peoples in the mid–twentieth century for Western tutelage and democracy. By 1945 flames of nationalism in Asia had incinerated the patience of colonized and exploited peoples. The Koreans, whatever their political views, demanded immediate independence, not occupation and trusteeship. An occupation was something to be visited upon conquered enemy lands, they argued. Korea had not been an enemy but a nation in the grip of an enemy, Japan. Nonetheless, occupation was forced upon the Koreans.

In the North the Soviets imposed central authority, which caused considerable hardship but, at the same time, brought land and social reforms avidly sought by the population. In the South the ferment and disorder of the Japanese surrender spawned political movements ranging from radical to reactionary, and their competition was often riotous. The Japanese collapse had come so suddenly that MacArthur, supreme commander in the Far East, had few troops adequately trained for such an occupation. He resorted to calling on forces stationed on Okinawa under the command of Lieutenant General John R. Hodge. Hodge, an infantry officer untrained in civil administration, landed in the midst of chaos. The occupation soon became a shambles. His forces were ill equipped to handle problems that were essentially political, social, and economic—a lesson the United States had to learn over again in Vietnam twenty years later.

In a familiar pose, the American military occupation forces favored conservative interests, not only because they seemed better qualified to manage the disrupted economy and help preserve law and order, but also because the rejection of leftist elements seemed a way to prevent Soviet influence from spreading into the South.

The events that led to war in 1950 were complicated and unpredictable. In the hostile atmosphere of the gathering Cold War, the United States and the Soviet Union could not come to terms on trusteeship. Border fighting and guerrilla skirmishes between the North and South Koreans kept tensions high. The Soviet Union heavily armed the North and withdrew its occupation force. The United States did the same in the South, but in a more roundabout manner. Because the reckless demobilization after World War II had left the United States with too few forces to confront new demands, especially in Europe, the army wanted to pull out of Korea. But the State Department objected on the grounds that a precipitate withdrawal would appear to be a Cold War defeat, weakening the United States in Asia and turning all of Korea into a Soviet satellite uncomfortably close to Japan. Truman sided with the State Department.

To pave the way for gradual withdrawal, however, the United States turned over to the United Nations the task of creating a permanent Republic of Korea below the thirty-eighth parallel. Such involvement of the United Nations, of course, ensured that its effectiveness as an organization would be undermined if its will were flouted and the South Korean government destroyed by aggression. (When the moment of crisis came, in fact, the preservation of the United Nations was a strong factor in Truman's decision to intervene.) The General Assembly approved a U.S.-sponsored resolution calling for U.N.-supervised legislative elections throughout Korea, followed by the establishment of a single national government and the end of the double occupation. The Soviets refused to permit elections in the North, but these took place in the South in 1948. The new national assembly drafted a constitution. Syngman Rhee, a Princeton graduate who headed a right-wing faction and was popular in Washington, was elected president of the republic, and the U.N. General Assembly declared his regime the only legitimate government in Korea. In the North, the Communists responded by forming the Democratic People's Republic of Korea, headed by Kim Il-sung, who had fought with Soviet forces during World War II.

The Americans armed the new government in the South, but not quite so heavily as the Soviets armed the North, because of a fear that Rhee might well attack North Korea if he became convinced he had superior power. By July 1949, the Soviets and Americans had withdrawn their occupation forces, leaving armed client states behind them. Thus Korea was divided between a Soviet-supported left-wing regime in the North and an American-backed right-wing one in the South, each eager to supplant the other and take over the entire peninsula.

America's formal, four-year occupation of South Korea had the effect of making the United States inescapably the protector of a new republic on the mainland of Asia. Automatically South Korea became linked with American security interests in the Far East, and the preservation of the Rhee government instantly became a sensitive issue in the Cold War. Ironically, before the withdrawal of the occupation force, the Joint Chiefs of Staff had held that Korea would be of no strategic value to the United States in a general war. As a symbol, however, Korea was another matter, a token of American purpose, resolve, and endurance at a time when neither side in the new global standoff was willing to back down anywhere.

Such were the circumstances when North Korean dictator Kim Il-sung, with the tacit approval of Moscow and Beijing, decided to start a war in a tinderbox. Writing in his memoirs about Kim Il-sung's discussions with Stalin, Khrushchev said, "[Stalin] was worried that the Americans would jump in, but we were inclined to think that if the war were fought swiftly—and Kim Il-sung was sure that it could be won quickly—then intervention by the USA could be avoided." But anyone who supposed in June 1950 that a great power would stand aside and allow its client state to be crushed by a rival was out of touch with the reality of the East-West struggle. Surely Kim was unaware of the mood prevailing in America, and he had no idea of the pressures that would descend on Truman the moment that North Korea's ninety thousand troops and one hundred and fifty Soviet-built tanks rolled across the thirty-eighth parallel.

A confluence of historical and political factors severely narrowed Truman's options. For five years the strains of the Cold War had been growing relentlessly. Soviet-American political clashes over Poland, the Balkans, and Iran; the disagreement between Washington and Moscow over the Baruch Plan to control nuclear weapons; the advent of the Truman Doctrine and the Marshall Plan; the Communist seizure of power in Czechoslovakia; the Soviet blockade of Berlin; and the signing of the North Atlantic Treaty—all made it painfully clear that the costs and sacrifices of World War II had failed to buy enduring peace. The momentous developments of 1949, when the Soviets ended the U.S. monopoly on atomic weapons and the Communists seized control of China, convinced many Americans that after the great victories over Hitler and Tojo, the West was losing out in the end to Stalin and Mao Zedong.

By such undertakings as the Marshall Plan, the Berlin airlift, and the formation of NATO, Truman had assumed great responsibilities for leading and protecting the noncommunist nations. To veer from that course and to shrink from a challenge, especially a military one, would be a dangerous show of weakness, it was widely believed in the West. To the leaders of Truman's generation, such logic was powerfully reinforced by the Allied experiences with Hitler in the 1930s. In their desperate effort to avoid a repetition of the ghastly events of 1914–18, the British and French had yielded to Hitler in the Rhineland and at Munich. Instead of being appeased, Hitler had been strengthened. He picked off his enemies one by one, faltering only at the English Channel and later in the Russian snows. When the North Koreans struck in 1950, Truman and all of his advisers tried to profit from what they understood to be the lesson—that, to quote Truman, "a breach of peace anywhere in the world threatens the entire world." As he told Congress in his first message on the Korean War on July 19, 1950, "The fateful events of the nineteen-thirties, when aggression unopposed bred more aggression and eventually war, were fresh in our memories."

Truman's experience in 1947, when Greece and Turkey seemed to be threatened by the Soviet Union and he had stepped in with American aid, contributed to the president's conviction that the right time to halt aggression was the first time. Whether the Soviets ever intended to take over Greece and Turkey in the first place is not clear, but those countries did remain free of external domination after a presidential speech enunciating what came to be known as the Truman Doctrine. Truman was inclined to perceive cause and effect. Indeed, on the very day he made his initial commitment of air and naval power to Korea, the president told George M. Elsey, an administrative assistant:

> Korea is the Greece of the Far East. If we are tough enough now, if we stand up to them like we did in Greece three years ago, they [the Soviets] won't take any next steps. But if we just stand by, they'll move into Iran and they'll take over the Middle East. There is no telling what they'll do, if we don't put up a fight now.

Despite the open-ended language Truman employed in announcing his doctrine, it was never intended to be applied in every crisis around the world; but as the president's comment to Elsey suggests, it seemed to inspire U.S. actions in Korea.

The ratification of the North Atlantic Treaty in 1949 was also a factor. The signatories had agreed that an attack on any one of them was to be considered an attack on all. If, therefore, the Soviets were to march westward, as many feared they might, the United States would be bound to go to war in defense of Western

Europe. There was widespread concern among Europeans that, in the final analysis, the Americans would not risk the nuclear destruction of their own cities by retaliating against the Soviet Union for attacks on Paris or Brussels, for example. When South Korea was invaded, the question arose whether this might be the forerunner of a broad Communist assault, with Western Europe as an ultimate objective. A second, but almost more important, question was whether the United States would let it happen.

The first State Department intelligence assessment after the North Korean attack was that, if successful, it would cause "significant damage to U.S. prestige in Western Europe. The capacity of a small Soviet satellite to engage in a military adventure challenging, as many Europeans will see it, the might and will of the U.S., can only lead to serious questioning of that might and will." Early press reports confirmed a suspicion in Europe that the United States would recoil from military challenge, thereby encouraging the Soviets to make trouble elsewhere. The British historian Hugh Brogan later wrote that if the United States had stood aside in Korea, that "would have put both Japan and Western Europe at risk, and would so have undermined the credibility of the United States guarantee as to weaken its ability to meet that risk."[3] The issue of credibility may have been overstated, but it was taken seriously by Truman and Acheson—as it would be later by President Johnson and Secretary of State Dean Rusk, when it came to defending South Vietnam. Although both wars demonstrated that the American people eventually tire of expending men and money in distant conflicts in order to prove that the United States will, if necessary, honor its treaty obligations somewhere else, in each case it took years to make the point.

A particularly potent fusion of the Cold War and partisan politics was bedeviling Truman by the eve of the North Korean attack. His upset of Thomas E. Dewey in the 1948 election left the Republicans in a nasty mood. All the issues they had used for twenty years had failed to unseat the Democrats. Just as Truman was savoring his victory, however, the tottering of Nationalist China under the blows of Mao's armies presented a spurious but powerful political weapon to conservative Republicans and their allies in the China Lobby: Truman had "lost" China, standing by while a monolithic alliance of the Chinese Communists and the Soviets rolled across the world landscape. Even as Nationalist China was staggering, Alger Hiss was indicted for perjury after denying that before leaving the State Department in 1946 he had passed classified documents to a communist agent. As a result of this sensation, the Republican case against Truman became more shrill: He had not merely lost China, but had done so as a result of alleged Communist machinations in the State Department.

Into this imbroglio lumbered Joseph R. McCarthy, a grinning, ruthless, irresponsible Republican senator from Wisconsin, who claimed in a speech in Wheeling, West Virginia, in February 1950 to have the names of 57 (some in his audience heard the figure 205) Communists working in the State Department. The uproar that followed lasted well into the Eisenhower administration and pushed the country into some of its darkest days of modern times. Anti-communism became an obsession, distorting politics and creating a witch hunt in the government to uncover alleged cases of disloyalty. McCarthy did not particularly seize on the issue of Korea. Rather he focused mostly on closed "loyalty board" cases that were an old story in the government but sensationally new to the public. But the effect of his onslaught was to put the Truman administration on the defensive on the communist issue. With congressional elections looming in November 1950, it would have been a grievous blow for the Democrats if Truman had permitted the "loss" of Korea on top of the "loss" of China.

Truman was outraged by the North Korean attack. The thought of letting the communist forces get away with it probably never crossed his mind. He was spending the weekend with his family at their home in Independence, Missouri, when Dean Acheson telephoned about the invasion. Margaret Truman noted in her diary that "we are going to fight." To the officials who met him on his return to Washington, the president said, "By God, I am going to let them have it!"

Truman chose a popular political course when he rebuffed the North Korean Communists, a consideration that was surely not lost on him. At the core, however, the decisive factors were strategic: "drawing the line," keeping an upper hand in the Cold War, reversing the precedent of Munich by halting aggression to head off a still larger war, safeguarding Japan, reassuring Europe, backing an ally, and standing by the United Nations.

Nevertheless, as General of the Army Omar N. Bradley, then chairman of the Joint Chiefs of Staff, later testified, the administration did not know what it was getting into. The armed services were not prepared for a major war in Asia. Even in responding to the North Korean attack, Truman did not believe he was entering a large engagement; he supposed he was up against a "bandit raid" that could be suppressed quickly, and so he decided not to bother seeking a declaration of war from Congress. No one had informed Truman that an urgent and dangerous situation was building up along the thirty-eighth parallel in early June, because American intelligence assessments were faulty. The mobilization in the North had been spotted, but through various rationalizations it was not interpreted as preparation for war. Perhaps the North Koreans would have reconsidered at the last minute if an alert American government had raised a cry of danger in the United Nations, rushed the Seventh Fleet to the Sea of Japan, put fighters and bombers in South Korea, and visibly reinforced MacArthur's four divisions in Japan.

As it was, the invaders swiftly ripped through the South Korean lines in a thrust so powerful that MacArthur's forces, once committed, were driven almost into the sea. They had to wage a desperate summer-long defense of their last stronghold around the southeastern port of Pusan ("the Pusan Perimeter") to prevent the complete conquest of South Korea. Truman increased American strength to eight divisions, which were gradually supplemented by units of fifteen other members of the United Nations, after the Security Council had for the first time voted to take collective action. The defensive battle drew the bulk of the North Korean army deep into the South. On September 15, 1950, in a daring amphibious landing behind the northern lines at Inchon, on the west coast near the thirty-eighth parallel, MacArthur trapped a large part of the enemy forces. The triumph at Inchon enabled the U.N. forces quickly to drive to the thirty-eighth parallel, thereby expelling the invaders from South Korea.

Nearly everything about the Korean War was wretched, including the fact that this fulfillment of America's original war aim —this triumph for collective security—was not regarded as an adequate victory, an acceptable end to the fighting. In the view of the responsible officials, the politicians, and the public, the return to the thirty-eighth parallel was a mere repositioning from which one last and seemingly easy offensive could be launched to "finish the job." Whether at the United Nations, in Washington, London, Rhee's headquarters, or MacArthur's command, the idea of halting at the parallel was unsatisfying. Rhee said that South Korean troops would advance into the North, even if all of the other forces stopped at the parallel. He played to a Western vision of ultimate victory, of deposing the communist regime in Pyongyang and clearing the way for the United Nations to reunify Korea once and for all, thereby causing Soviet influence to decline in the Far East.

Misconceptions linger about MacArthur's role at this point in the war. The belief persists that in the desire to score a great victory, he personally ordered the forces under his command to cross the parallel. In fact, MacArthur was instructed by President Truman to cross the parallel and destroy what was left of the North Korean army. Secretary of State Acheson and Secretary of Defense George C. Marshall agreed with those instructions, which were sent through the Joint Chiefs of Staff.

MacArthur was in every sense a powerful figure. He was an acknowledged war hero. His handling of the Japanese occupation was widely considered masterful. Having absented himself from the United States for fourteen years, he was untarnished by the wear and tear of domestic politics. Indeed, remoteness enhanced his aura, lifting him in the public mind above those battling in the trenches of Washington, like Harry Truman. MacArthur had wealthy and powerful friends at home and ardent support from the publishing empire of Henry R. Luce and the Hearst and Scripps-Howard newspaper chains. A righteous orator and aristocratic Republican who was contemptuous of the New Deal, MacArthur was for many the embodiment of duty and honor, of anti-communism and American military might. He was disliked by liberals but had long been an idol of GOP warhorses weary from trying to find someone who could defeat Roosevelt. Truman himself suspected that at a propitious moment MacArthur might well make a triumphal return to the United States and seek the Republican presidential nomination.

Since early in the Korean War MacArthur had wanted to cross the thirty-eighth parallel and end communist rule in the North. His friends in Congress and the press supported his ambition. For Truman to have denied him this chance would have been difficult and politically treacherous. A halt when the enemy seemed to be on its last legs would have been seized upon by Republicans as the ultimate proof that the administration was "soft" on communism. In the forthcoming presidential campaign of 1952, Republicans would have charged that Truman and the Democrats had lost their nerve and thrown away ultimate victory in Korea.

The fact is that until after the parallel had been crossed, MacArthur and Truman had no fundamental difference on Korean strategy, except that Truman had vetoed the hot pursuit of enemy planes into Manchuria by American aircrift. Officials in the Pentagon and the State Department were talking about the need to cross the parallel just as early as MacArthur. Truman and his advisers saw as well as the general did the strategic, political, and diplomatic advantages of ending communist rule in the North. That was why on September 27, 1950, Truman instructed him to advance.

The decision was a disaster, because of Chinese military intervention. As for what would have happened if Truman had held MacArthur at the parallel, one can only speculate. Korea would, of course, have remained divided, as it is today. The United States would have had to announce its intention to retain forces indefinitely at the thirty-eighth parallel, a policy that would have caused tremendous political bitterness at home at the time. (The fact that U.S. troops did remain in Korea—and remain there today—had diminishing political impact over time, as the American public began to pay less attention to that distant peninsula; even those who supported their continued presence never imagined they would stay so long.) By not crossing the parallel, Truman would have avoided armed conflict with the Chinese, but the Chinese would have been hostile to the continuing American military presence in South Korea anyway. Washington's relations with Rhee would probably have been all but unmanageable. Nevertheless, in retrospect, almost anything would have been better than the consequences of the decision to cross the parallel.

Frequently after the Inchon landing the Chinese government had declared, directly and indirectly, that it would oppose any attempt by the United States to crush China's new communist neighbor on the Korean peninsula, which would have meant bringing American power to the very boundary of Manchuria at the Yalu River. The threats worried Truman. He asked the Central Intelligence Agency for a special assessment. Consistent with opinion in other quarters of the government, the CIA replied that, barring a Soviet decision for global war, no intervention by China in North Korea was "probable" in 1950. Though still uneasy, Truman accepted the conclusion.

In the wake of these miscalculations in Washington, which matched the findings of his own sorry intelligence operation, MacArthur launched his reckless "win the war" offensive toward the

Yalu on November 24, 1950. Three hundred thousand Chinese troops, concealed in the mountains and ravines south of the river, hurled themselves at MacArthur's forces, so nearly trapping the U.S. Tenth Corps against the Sea of Japan as to cause a spate of talk in the Pentagon and the State Department about the possible use of nuclear weapons. After harsh battles the U.S. Eighth Army, the main strength of the U.N. ground forces, retreated to South Korea and finally stabilized a new dividing line, again roughly along the thirty-eighth parallel.

The Truman administration thereupon abandoned the idea of reunifying Korea, a decision that was anathema to MacArthur, who was determined not to end his career in stalemate. He proposed that Truman treat the Chinese intervention as a new war. It was then that the general split with the administration, urging that the United States bomb and blockade China, issuing statements without obeying Truman's explicit instructions to clear them in Washington, deliberately thwarting an administration effort to get a cease-fire with China, and going behind the president's back with a letter to Joseph W. Martin, Jr., Republican leader of the House of Representatives. Truman relieved MacArthur of his command, unleashing a profound political crisis.

Under MacArthur's successor, Lieut. Gen. Matthew B. Ridgway, the U.N. forces regrouped, attacked and, with superior air power against extended enemy supply lines, inflicted such severe casualties on the Chinese troops that the climate for negotiations became unmistakable. Truce talks began on July 10, 1951, but for the remainder of Truman's term they were deadlocked, because he would not agree to the involuntary repatriation of communist prisoners of war held by U.N. troops; he would not send back captured North Koreans who did not want to go home because they feared they might be killed or tortured by Kim Il-sung's regime. The battering years after World War II had already eroded Truman's leadership; the Korean War tore apart much of what was left of it. In 1953, a truce line running roughly along the thirty-eighth parallel was accepted by President Eisenhower. It still divides North and South Korea today.

The Korean War was a conflict for which no one in the world seemed to feel any enthusiasm. Europeans saw it as an unwanted diversion from the tasks of recovery from World War II. It brought serious new financial burdens to a world system that was still reeling from the previous round of fighting. Korea also provided a new area of East-West conflict in the United Nations, and support for U.N. intervention on South Korea's side caused domestic political strains in many member nations. Above all, there was widespread international concern that the fighting in Korea might get out of hand and expand into a general war.

It was that concern which caused the most trouble for the Truman administration, because it threatened to disrupt continuing support for the war effort in the United Nations. Underlying this threat was a growing fear among America's allies that MacArthur might run amok and somehow extend the effort in Korea into military support for Chiang Kai-shek and his Nationalist forces on Taiwan against the communist government of mainland China.

The traditional allies of the United States were frankly relieved by Truman's initial decision to intervene against the North Korean invasion. Cheers erupted in the House of Commons in London when the news came. Trygve Lie, the influential Norwegian secretary general of the United Nations, enthusiastically helped line up U.N. assistance in the fight against North Korea. The United States then was the dominant influence in the United Nations, and at American urging the Security Council voted first to condemn the breach of the peace and then to declare sanctions against North Korea. The votes were nine to zero and seven to one, respectively. (Yugoslavia, then having severe problems of its own in trying to establish its independence from the Soviet Union, cast the sole vote against sanctions.) The Soviet delegate was then boycotting the United Nations over its refusal to transfer China's seat to the communist regime in Peking, and so he did not even attend the sessions, let alone cast a veto.

Fifteen other nations agreed to send military units to Korea, most of them too small to be effective. The British immediately placed at MacArthur's disposal its warships and crews in Asian waters, and later sent some forty thousand troops from around the Commonwealth; but the French contribution to the war effort was modest. Acheson explained to Truman at the outset that little help could be expected from the French, who already had their hands full in Indochina. While Truman wanted all the help he could get and thought he would get much more than was offered, the record does not show that he put pressure on allied countries for greater assistance, as Lyndon B. Johnson did later when he urged the Asian and Pacific nations that supported the United States in Vietnam to expand their forces in the field. If, in the Korean War, any of the allied participants felt happy to be represented on the battlefield, it is one of the great untold stories of the era. On the other hand, if any of them felt unjustly used by the United States, none ever showed it by symbolic withdrawal. In general, the allies seemed to share the attitudes of the American people toward the Korean War: When a communist army struck, they willingly supported the policy of drawing the line. When the going got rough, however —when the war became drawn out and, above all, when China intervened and threw MacArthur back—they were shocked and demanded negotiations.

Time and again the allies felt frightened, often unduly so, over the purposes and actions of the U.N. commander. Among the first to be alarmed by MacArthur were the British, especially when he turned up on Taiwan to confer with Chiang. Later, MacArthur issued a statement that could be read as a hint of impending military support for the Nationalists, an interpretation he denied. Soon afterward, he sought Truman's approval for hot pursuit of enemy fighter planes across the Yalu into Manchuria. The aroused allies protested, and Truman denied MacArthur's request, much to the general's wrath. He did not relish having other nations meddle in his planning on their behalf.

As for the fateful crossing of the thirty-eighth parallel, the allies also reacted much as the Americans did: in effect, "go to it, get rid of the Korean problem once and for all." But when MacArthur did cross and the Chinese staged their surprise response, then everybody thought differently. The allies were terrified that MacArthur might bomb China and plunge the entire world into war. As if to compound the problem, at a press conference on November 30, 1950, Truman answered some questions clumsily, creating a false impression that he was considering using atomic bombs in Korea. Such was the panic in Europe that British Prime Minister Clement R. Attlee flew to Washington on a mission to prevent a third world war. Five days of talks with Truman calmed him, and he returned to London satisfied with what he had learned.[4]

The Korean War had important implications for the United States in places far removed from the battlefield. Korea led Truman into several cardinal decisions of global import, decisions that added a military dimension to the Cold War and affected the balance of power in the world.

At a critical time during the fighting he approved National Security Council memorandum 68 (NSC 68), a policy paper prepared early in 1950 by an interdepartmental group headed by Paul H. Nitze and working under Acheson's eye. Undertaken largely as a result of the Soviets' success in breaking the American nuclear monopoly, it recommended a huge program to expand U.S. military might, conventional and nuclear, and to rearm America's allies. Truman might never have approved such a document were it not for the Korean War. Once he did, however, the total budget for defense and international affairs rose from $17.7 billion in fiscal year 1950 to $53.4 billion in fiscal year 1951.

The startling ease with which the North Koreans had overrun Rhee's troops aroused Truman's and Acheson's fears of what might happen in Europe if the Red Army were to strike through Germany. Only a few years after Adolf Hitler's death, the worsening Cold War had spurred allied talk about rearming West Germany at some vague point in the future. The Korean War ended the vagueness. As one way of strengthening NATO, Truman authorized Acheson to negotiate with the British and French about German rearmament. French concerns slowed the process, but the German military began to be rebuilt during the Eisenhower administration. As another, and more immediate, way of putting steel in NATO, Truman, after a tilt with the remnants of the isolationist bloc in the Senate, added four additional American divisions to the two already on occupation duty in Europe. The permanent stationing of American troops overseas at a time of peace was without precedent in American history.

While the war in Asia thus wrought profound changes in Europe and in America's relations there, the sorriest legacy for the United States still lay in the Far East. It was not only that the Korean War was cruel, costly, and indecisive, but also that it led indirectly to the far worse disaster of American involvement in Vietnam.

The Truman administration had trouble from the outset with the Chinese Communists, because Truman had inherited from Roosevelt a policy that recognized Chiang's regime as the legitimate government of China. In trying to save Chiang with economic and military assistance, Truman ran afoul of Mao Zedong. The American people hoped that the Communists would not win the civil war. When they did, Truman suffered grievously for it, and the most unfortunate way his administration reacted was by bolstering the French in Indochina, out of a concern that the Chinese might try to extend their sway to Southeast Asia. The course Truman adopted early in 1950 was to recognize a French puppet regime in Vietnam under Bao Dai and grant it small-scale aid. A few months later, when the Korean War began and fear of communist expansionism redoubled, Truman escalated the financial assistance and sent a military mission to Vietnam, a mission that was to be steadily enlarged by Dwight D. Eisenhower and John F. Kennedy until, finally, Lyndon Johnson went the full course and sent combat troops.

Despite the bad blood between Washington and Peking, the Truman administration was moving toward formal, if initially cool, relations with China in the late 1940s, a development that might have altered the history of the years after 1950. But after war erupted in Korea, MacArthur crossed the thirty-eighth parallel and Chinese troops massively attacked American soldiers, the United States and China became totally estranged for two decades. The Chinese onslaught, which denied the United States ultimate victory in Korea, had a profound effect on the American people, not to mention Eisenhower, Kennedy, and Johnson. China became an enemy more feared in the United States than even the Soviet Union. Anxiety over the possibility that China would move southward into Vietnam and beyond was a major factor behind the determination of those three presidents to prevent the Communist conquest of South Vietnam.

The image of a limited success in Korea, where the United States had at least preserved the South for an astoundingly prosperous future outside the communist orbit, lingered as Americans plotted strategy in Vietnam: To save the South Vietnamese regime of Ngo Dinh Diem, the communist-led National Liberation Front in South Vietnam would have to be crushed. A line would be held along the seventeenth parallel, just as it had been across the thirty-eighth parallel in Korea, in order to prevent the North Vietnamese army from infiltrating to attack the South Vietnamese. American aid would flow, and South Vietnam would survive as a barrier to communism, as South Korea had. Unfortunately, the

Americans had failed in Korea to grasp the intractable problems posed by militant nationalism, and nationalism in a more powerful form thwarted them in Vietnam.

Looking back on Korea, one remembers, above all, frustration. No American, from the president on down, wanted to go to war. The United States had other, more important business at hand. If in 1945 Truman had stayed clear of a Korean occupation, perhaps limiting American involvement to a temporary military mission to accept the surrender of some of the Japanese troops on Korean soil, hostilities might have been avoided in 1950 because Korea would not have become a flammable issue in the Cold War. Instead, the United States did become deeply involved, and Korea was a vital Cold War issue by the time Kim Il-sung attacked. If American intervention at that point was not inevitable, it was, given the circumstances, unimaginable that the Truman administration would simply tolerate the invasion. The period was an unhappy one for the United States, made no easier by strains with its allies. In the end, however, the Korean War was not a great turning point in American history. It did not demonstrate the folly of a global policy of containment. Otherwise the United States would not have been involved in the calamity of Vietnam, which dims the agony of Korea.

4 Confronting a Revolutionary Legacy

DONALD F. McHENRY

Donald F. McHenry has had a distinguished career in international diplomacy. After serving as deputy U.S. representative to the United Nations from 1977 to 1979, he was U.N. ambassador for the balance of the Carter administration. Since 1981 he has been university research professor of diplomacy and international affairs at Georgetown University; he is also president of International Relations Consultants and a director of several American corporations. Mr. McHenry's writings on foreign policy have appeared in many periodicals and journals, and he is the author of a book on U.S. policy toward its semiautonomous trust territories, *Micronesia: Trust Betrayed*.

IN THE AFTERMATH of World War II, the Western world was faced with a new revolutionary circumstance: the assertion by some 600 million people, more than a quarter of the world's population at the time, of the right to form their own governments. Since then, millions more people, including many in those same nations that had just been established, have decided, not always in a peaceful or orderly way, to cast off their existing governments and begin anew. As a result of these momentous events, the cast of characters in international affairs has changed permanently.

The United States has long since forgotten its own revolutionary moment, in 1776, when the original thirteen colonies challenged the established international order and demanded a government deriving "just powers from the consent of the governed." Indeed, the United States—particularly in the period since World War II—has now come to symbolize the status quo; it is the established power that finally must learn how to deal with the successors to its tradition as an upstart nation. This has been all the more complicated because many of the newer nations can trace their own cultural and political traditions back several centuries, and they proudly refuse to accept the idea that outsiders know what is best for them. The new nations have different perspectives and priorities, different cultures and experiences, with which the United States copes poorly, if at all. But cope it must, while maintaining its own economic and political interests and reaching a sensible definition of its international security responsibilities.

While the political map of the world was changed permanently in the postwar period, it was not changed finally. Indeed, it is still in a state of flux. Uncertainty and instability are inherent in the evolution of new nations and new institutions, and even in the efforts of older nations to adapt their form of government. If Europe and the United States are any guide, instability in the new countries will last for generations to come. How the United States has reacted to these developments so far tells a great deal about American foreign policy today. Not only has there been considerable tension between the United States and many of the new nations, but also, as a result, between the United States and friendly governments with which Washington had longstanding, if sometimes shaky, relations. Moreover, these changes in the shape of the world have led to new tension in relations between the United States and the Soviet Union. For example, whenever one of them spots an ally or potential ally among the new countries moving toward the opposing camp, whenever nationalist groups or established governments seek to enhance their positions by appealing for help from one of the superpowers, or whenever the Americans or Soviets or a regional antagonist sees an opportunity to enhance its regional position by making mischief, Americans seem to feel threatened by such jockeying for advantage in the Third World. Opportunistic rhetoric is often taken at face value, and the result is a widespread belief in the United States that the new countries are anti-American and probably pro-Soviet.

The issue is of more than passing interest and fraught with significance for the future. Although it is unlikely that in the nuclear age the United States and the Soviet Union would go to war in Europe or the Far East, where their real interests clash, or even consciously in the Third World, there remains a great danger that they may back into conflict in the developing countries, because they attach greater importance than their vital interests require to disruptions in unstable areas of the globe. If America seeks to avert that danger, it must arrive at a better understanding of the younger nations than it has today, and it must recognize the shambles of the 1950s and 1960s, when the most dramatic changes took place.

The United States approached the post–World War II period and its new responsibilities with some attitudes well ingrained, and that should have boded well for the newly emerging countries. In theory at least, the American belief in the right of people to govern themselves, and to insist that government remain responsive to them, is stated clearly in the Declaration of Independence. The young American nation went to war on behalf of those principles in the late eighteenth century, and it fought a bloody fratricidal war in the 1860s over the issue of how broadly to apply and interpret the principles.

There had been many prior revolutions in the world, also accompanied by noble statements of purpose, and there would be many more to come, but the Declaration of Independence is rivaled by few documents of its kind in its reach and its impact. The United States, after all, was the first of the new nations to break away from its foreign rulers. Borrowing heavily from John Locke and from traditional writings on natural law, the rebels gathered in Philadalphia provided a philosophical foundation that would be used more than two centuries later to revolutionize the world. The declaration boldly asserted that a people could, when necessary, break their political ties with another and "assume among the Powers of the earth, the separate and equal station to which the Laws of Nature and of Nature's God entitle them." Truths were not subject to debate for these people fed up with colonialism; they were "self-evident":

> . . . that all men are created equal, that they are endowed by their Creator with certain unalienable Rights, that among these are Life, Liberty, and the pursuit of Happiness. That to secure these rights, Governments are instituted among Men, deriving their just powers from the consent of the governed, that whenever any Form of Government becomes destructive of these ends, it is the Right of the people to alter or to abolish it, and to institute new Government, laying its foundation on such principles and organizing its powers in such form, as to them shall seem most likely to effect their Safety and Happiness.

To be sure, the Declaration of Independence argued that actions to change political ties with another government or to change that government itself were not to be taken lightly. Indeed, experience had shown that humankind was more disposed to endure suffering than to abolish an existing government. "But when a long train of abuses and usurpations, pursuing invariably the same Object, evinces a design to reduce them under absolute Despotism, it is their right, it is their duty, to throw off such Government, and to provide new Guards for their future security," asserted the document.

The American Revolution became the reference point for nationalists the world over as they too sought to assert equality among persons; their right to life, liberty, and the pursuit of happiness; their right to govern themselves; and, their ultimate right, indeed duty, to change their government, by force if necessary, to ensure that it become responsive to them.

From the outset, of course, there was an incongruence between the idealism of the declaration and American practice. More than half of the southern signatories who solemnly declared that all men are created equal were themselves slave owners, and the new Constitution would soon enshrine a unique American brand of inequality in Article I, Section II, counting a black person as three-fifths of a white. Moreover, in the establishment of the colonies themselves and in the expansion of the United States westward, the American settlers dealt unfairly and often ruthlessly with the Indian population. Frequently, treaties were entered into but were unilaterally broken when that served the new "national interest." Much of the justification for this westward movement was based on America's "manifest destiny," but it did not stop with the natural boundary of the Pacific Ocean. In 1867 the United States purchased Alaska from Russia. In 1898 it formally annexed the Hawaiian Islands, having declared them an American "protectorate" five years earlier, when Washington intervened in sympathy with a revolt against Hawaii's queen. The United States took the Philippines, Puerto Rico, Cuba, and Gaum from Spain, ostensibly in response to the bombing of the battleship *Maine* in Havana harbor in 1898 but also, and just as important, in response to the opposition of Cuban exiles and American investors to the backward policies of the Spanish government in Cuba. The Virgin Islands were bought from Denmark in 1917. In 1899 a long-simmering dispute between the United States, Germany, and the United Kingdom over the Samoan Islands in the Pacific ended with their division between the United States and Germany, while Britain got concessions in Africa.

America did not limit itself to the acquisition of territory in its first century, or its second. As the battle ribbons of the U.S. Marine Corps colorfully attest, the United States has frequently intervened, usually in Latin America, to put down a revolt or to shore up a favored government.

From the outset, American "imperialism" differed in important respects from that of Europe. The United States, despite a strong desire to influence events in its own hemisphere and occasionally beyond, never evinced an interest in developing a colonial empire. Relatively few Americans went out to settle these newly acquired territories the way Europeans did theirs, and U.S. administrators there professed to have benign motives and temporary missions. Given their own history, Europeans may be forgiven their skepticism if they doubted that the Americans, however much they justified their involvement as a "civilizing" influence, would actually promote government by the inhabitants themselves. However, on the whole the United States did provide for self-government for territories under its control, if not always for self-determination.

The government of Cuba was turned over to the Cuban people in 1902 after the restoration of order, a considerable step even if, under the terms of the Platt Amendment, the United States intervened again several times during the next two decades, whenever it did not approve of the course of events in Havana. Similarly, after several false starts, the Philippines were put on the road to self-government in 1913, and in 1934 Congress passed a law setting Philippine independence for 1945, after ten years of special U.S. tutelage and transition. Alaska and the Hawaiian Islands were ultimately integrated into the United States in such a way as to make no distinction between them and the other states, or between their citizens and citizens within the forty-eight contiguous states. Their eventual status was heavily influenced by the nature of their relationship with the United States in the century before their statehood. One might speculate that they might have been candidates for separate nationhood if that relationship had been at all strained.

Only Puerto Rico, Gaum, the Virgin Islands, and American Samoa remain in some form of territorial status today, although each is self-governing. Of these, only the "commonwealth" of Puerto Rico has had a significant growth of nationalism and has an uncertain future. In the others, a long and close association with the American system has probably resulted in a greater desire for integration into the United States than for the attainment of independent status.

American territorial acquisitions were limited for a variety of reasons. Most of the Western Hemisphere was already independent (and off limits to Europe under the Monroe Doctrine). The rest of the world had been rather thoroughly carved up; the major problem remaining was to assure access to markets like China. But more important was the debate within the United States. The acquisition of territory was sharply criticized as inconsistent with the tenets of the Declaration of Independence and the democratic way of life. When it happened, therefore, it was often the result of bold executive action, and whatever the real reason for intervening or seizing territory, it was rationalized in moral trems and accompanied by a commitment to turn over government to the people. In his letter accepting the Republican presidential nomination in 1900, William McKinley exalted his party's role in the abolition of slavery in the United States, and asserted that it would "not be guided in its conduct by one set of principles at home and another set in the new territory" overseas. According to one study of President Franklin D. Roosevelt's foreign policy, "the promotion of democracy, the ending of internecine violence and perpetual revolution, the establishment of order, the advancement of education and health and general welfare of the natives, and the raising of standards of living were invariably major objectives" of America's forays overseas.[1]

Except possibly in Puerto Rico, the United States never had to face the kinds of nationalist pressures that Europe would come to know in its colonies. This was an obvious advantage in the short run, but in the long run it probably made the United States naive about nationalism as a factor in international affairs.

The United States brought the same moralistic tone to its participation in the two world wars. President Woodrow Wilson advocated self-determination and denounced the territorial aggrandizement that had characterized previous conflicts; he felt that the spoils need not necessarily belong to the victor. In the covenant of the new League of Nations, Wilson introduced the concept of international oversight for territories surrendered by the losers, an oversight which he thought should end with the turning over to the subject peoples of responsibility for their own affairs. In the event, the ingrained habits of European and Asian powers led them to insist on doing business as usual, and Wilson was forced to accept some territorial deals as part of the Versailles peace treaty. Germany was not only stripped of its military power and forced to pay heavy reparations, but it was also required to surrender its colonies in Africa and Asia, to restore Alsace-Lorraine to France, and to cede portions of territory to Poland, Denmark, and Belgium. Nevertheless, some of the concepts insisted upon by Wilson, including self-determination and international oversight of the status of dependent people, were to have long-lasting effect. All were expanded in their application following World War II.

In his book *Franklin D. Roosevelt's World Order,* Willard Range traces how FDR, a one-time supporter of "humanitarian imperialism," became an "ardent anti-imperialist." Unlike Wilson, William Jennings Bryan and others who based their actions on humanitarian principles but did not hesitate to intervene in the Caribbean and Central America, Roosevelt's original support of imperialist gestures was combined with realism. "For realistic or geopolitical reasons he favored United States domination of whatever land or water was necessary to assure protection of the Panama Canal and the water approaches to the United States," Range says.

Sometime in the late 1920s Roosevelt came to believe that despite high objectives, the United States had reaped a "whirlwind of ill will." Shortly after he became president, he signed legislation (which his predecessor, Herbert Hoover, had once vetoed) providing for Philippine independence; he also opposed U.S. intervention in Cuba in 1933 and 1934, pulled the Marines out of Haiti, and, of course, pronounced the "Good Neighbor Policy."[2]

The evolution of Franklin Roosevelt's views toward European colonization can be traced in the remarkably candid off-the-record press conferences he was fond of holding, as well as in the writings of his son Elliot and in State Department documents on the planning for the United Nations. The president seemed particularly to have been impressed by the conditions he saw in the Gambia while en route to Casablanca for a wartime meeting in 1943. He accused the British of exploitation, of taking resources out of the country but putting little in, and of taking few steps to educate the people or prepare them to govern themselves:

> . . . we have got to realize that in a country like Gambia—and there are a lot of them down there—the people, who are in the overwhelming majority, have no possibility of self-government for a long time. But we have got to move, the way we did in the Philippines, to teach them self-government. That means education, it means sanitation, it means all those things.[3]

Roosevelt made similar observations about French rule in Indochina and Dutch rule in Java (Indonesia). For him, the end of the war would see the end of exploitation of one nation by another. He had little sympathy for British ideas of empire or for Charles de Gaulle's view that France could not recover her rightful role in world affairs if she lost her Asian and African territories. Roosevelt's idea was that the United Nations should send a committee to expose colonial powers who failed to "come up to scratch." He did not recoil at Churchill's suggestion that a similar committee be sent to the American South. "It would be a grand thing. . . . Why not?" he said. "We have got some things to be ashamed of, and other things that are not as bad as they are painted. It wouldn't hurt at all—bring it all out."[4]

American sentiments toward the nationalist movements that would grow in number and force after World War II were spelled out in August 1941 in the Atlantic Charter; there Roosevelt and Churchill, echoing Wilson's Fourteen Points, signaled the principles on which peace would be based. Again, there was a call for self-government and self-determination, coupled with a warning against territorial aggrandizement. However, just as they had at Versailles in 1919, the representatives of the Old and the New worlds used the same terminology to mean entirely different things. Churchill thought the Atlantic Charter applied to Europe, not the British colonies. When, later, in his most expressive way, the British leader stated that he had "not become the King's First Minister in order to preside over the liquidation of the British empire," Roosevelt dismissed Churchill's complaint as "mid-Victorian" ("dear old Winston will never learn on that point," FDR said) and insisted that the question of the future political status of Europe's colonies had been settled in the Atlantic Charter. The Roosevelt-Churchill relationship easily withstood their differences on colonialism, but not so the relations between Roosevelt and de Gaulle. French suspicions, wrote Tony Smith, left "a permanent mark on French attitudes toward the United States whenever colonial questions arose."[5]

Roosevelt recognized that a period of preparation was necessary before the colonies could actually become independent. He referred repeatedly to the American experience in the Philippines as a model, complete with early agreement on a definite date for independence. The whole process, he thought, could be conductd under the trusteeship arrangement that he envisioned as a part of the United Nations. Some form of international oversight, including periodic inspections by an international committee, he believed, would publicize the failure of the colonial powers to tutor their colonies properly so as to advance their educational, economic, and social development.

We do not know with what vigor and with what success Roosevelt might have pursued his ideas on decolonization had he lived. He certainly felt no inhibition about disagreeing publicly with Churchill, his closest wartime collaborator. And, of course, Roosevelt did not hide his differences with the proud and nationalistic de Gaulle. In the end, however, Roosevelt lived long enough to see that, even in Europe, his ideas on self-determination had begun to be circumvented—by another wartime ally, Joseph Stalin.

While not as far-reaching as it might have been, the degree of international oversight of colonies written into the U.N. Charter was a significant advance, and it was in no small measure due to American postwar planning. Of course, it would have been difficult not to bring about some reforms, given the other forces that were by then at work: With the independence movement close to ultimate victory in India, nationalists everywhere were asserting themselves. The entire colonial system had been disrupted during World War II, when the European powers were effectively cut off from their overseas possessions, and now the Europeans had to concentrate most of their economic resources and political efforts on rebuilding their own nations. Those few of the colonial subjects who had gone to Europe or the United States to study, fought in Allied battalions, or attained a greater degree of responsibility at home during the war tended to be less compliant with their masters; they knew what they wanted, and for them it was simply a question of how to obtain it. No matter how limited the colonial powers may have seen their obligation to promote change to be, the process had moved ahead significantly. The argument about whether to decolonize was largely over, even if some, like the Portuguese, would hold on grimly for another thirty years. The new arguments were over more practical issues, but each new example of independence increased the pressure for similar moves elsewhere.

Having provided much of the philosophical foundation and political pressure for the movement toward self-determination, for the belief that effective government must be based on the consent of the governed, and having portrayed itself as a moral force in international affairs, how did the United States soon come to be regarded as a lukewarm supporter of decolonization and a defender of established and often unrepresentative governments facing challenges from their people? The answer can be found in the always difficult circumstances which the United States and its allies faced after the war, and in a growing American ambivalence about how to exercise its new power in the postwar world.

From the start, the provisions of the new United Nations Charter dealing with decolonization represented a major compromise with the colonial powers. The trusteeship system, for example, which was spelled out after Roosevelt's death, was far more limited than he had envisaged. It covered only eleven countries, mostly those just taken from defeated Axis powers, plus others that had been taken from defeated powers after World War I and placed under the mandate system established by the League of Nations. These eleven were subject to the kind of international supervision that Roosevelt had in mind, and their administrators were pledged to move them toward self-government or independence, whichever the people desired. As for the bulk of the world's colonies at the time, there was no obligation to place them under trusteeship. Under Chapter XI of the United Nations Charter, there was only a vague, general duty to promote self-determination and the social, educational, economic, and political advancement of the people. Significantly, however, no visiting missions were required; there was no obligation to report to the United Nations on political advancement; and self-government, as distinct from independence, was the stated goal toward which the colonial powers were supposed to bring their possessions.

Among the compromises reached were these:

- For Roosevelt, trusteeship was a sweeping concept that would provide the basis of international oversight for Greenland, Antarctica, Easter Island, Galapagos, Indochina, Micronesia, and other spots on the globe with an uncertain future. But of that list, only Micronesia became a trust territory; and even there it was left to Harry Truman, as one of his first presidential acts, to resolve a dispute between the military and the State Department as to whether U.S. security interests in the Pacific required annexation or a special trusteeship arrangement.
- An attempt by the U.N. General Assembly to dispatch visiting missions to a variety of colonies found the United States arguing with Britain and France over whether such missions could be sent without the consent of the colonial power. Indeed, confronted with a request to apply its principles to itself, the United States was at first reluctant to receive visiting missions in the Virgin Islands, Guam, and Samoa.
- Roosevelt had been fond of citing the advance timetable under which the Philippines gained independence, but no such requirement was included in the charter. In the United Nations itself, the United States ultimately opposed a timetable requirement, arguing that this was the prerogative of the colonial power in each case.
- Roosevelt believed that the principle of self-determination was applicable to any national group, no matter how small. He was confident that they would be responsible and protect their own best interests if allowed to act in an informed manner in an atmosphere of calm. But in the United Nations, the United States, in response to a British entreaty, abstained

> on the key Colonialism Declaration, in part because it applied to numerous small territories and seemed to modify the original charter to favor independence as the ultimate outcome of self-determination.

None of this is to suggest that Roosevelt, had he lived, would not have had to modify his views and accept compromises. He had already done so with respect to Poland, a place of great American concern, when he accepted less than full assurances of its sovereignty at Yalta. Clearly, Roosevelt favored a period of tutelage before colonies became independent. He probably would have agreed with the provision of the Colonialism Declaration which said that inadequacy of political, economic, social, or educational preparedness should never serve as a pretext for delaying independence; but he did regard these as legitimate considerations that might affect the timing of independence. In any event, Roosevelt does not seem to have envisioned some of the practical problems that would arise in the decolonization process—some of which arose before his death. And he did not address the question of what would happen if a friendly but repressive government was overthrown by its own people and, worse still, introduced alien social, political, and economic concepts into the Western Hemisphere. These circumstances and other complex, troublesome events were to come about repeatedly during the forty years after the war, inevitably coloring the American perception of nationalism and the nationalists' perception of America.

American principles became very difficult to apply in practice, especially given the new perception of U.S. responsibilities as a superpower. Although the United States was clearly on record as favoring strict international oversight of colonies, it found itself supporting the right of the colonial powers to resist such requirements. Washington argued that while the Europeans should not be required to submit information on political developments in their colonies, they ought to do so voluntarily, as the United States did for territories under its control. Similarly, while the United States was on record favoring timetables for independence, it opposed efforts by the General Assembly to set them. It favored the dispatch of visiting missions to colonies only if the mission was invited by the colonial power. And with regard to the Portuguese territories in Africa, where the circumstances were especially egregious, the United States argued that the United Nations could not declare that they were non-self-governing, but would have to content itself with establishing criteria that the colonial power could use in deciding on its own whether a colony was self-governing. The United States knew, and stated, that the Portuguese territories were not really governing themselves, but it wanted Portugal to make this admission against its own interest before anything was done, because the General Assembly was, in Western Europe's (and now America's) view, not entitled to make such a finding.

These legalistic approaches invariably made the United States seem to be trying to play both sides against the middle, and American exhortations in favor of self-determination inevitably came to sound hollow. With regard to the Portuguese territories, for example, the United States provided educational and other assistance to African nationalist movements while at the same time defending Portugal's cause in the United Nations and cooperating with Portugal in NATO in ways that freed the authorities in Lisbon to strengthen their military efforts to retain their colonies. The result was not to please, but to antagonize, both sides. The Portuguese believed that the United States was insufficiently supportive of a NATO ally, while the nationalists came to look upon the United States as the principal pillar in Portugal's efforts to perpetuate its repressive rule over African people. (It is no surprise, under the circumstances, that the nationalists gradually and bitterly turned elsewhere.)

Over and over again, the United States was caught in a similar bind, between its support for the principle of self-determination and its loyalty to an ally. President Dwight Eisenhower's refusal to support the cause of Algerian independence in the United Nations, or even to exert political pressure on France, was interpreted by Algerian nationalists as an unstinting endorsement of the French position that ignored the increasingly violent nature of the conflict in Algeria. Symbolic as it was of the worldwide stirrings of nationalism, the Algerian issue was picked up by American liberals as an opportunity to redeem the U.S. reputation. Even today, the dramatic floor speech in 1957 in support of Algerian independence by Senator John F. Kennedy is remembered in the developing world as an evocation of true American principles. For Kennedy, the Algerian situation was "critically outstanding"—a "clash between independence and imperialism."[6] On the other hand, America's last-minute abstention on the Colonialism Declaration in the United Nations in December 1960, followed by continuing disputes between the majority of U.N. members and the United States over its implementation, had a lasting impact. For today's generation of Third World nationalists—and, for that matter, for idealistic young Americans—it is hard to remember the important role the United States played in the decolonization movement.

As Kennedy put it in his landmark Senate speech on Algeria, the U.S. policy was characterized by

> tepid encouragement and moralizations to both sides, cautious neutrality on all real issues, and a restatement of our obvious dependence upon our European friends, our obvious dedication nevertheless to the principles of self-determination, and our obvious desire not to become involved. We have deceived ourselves into believing that we have thus pleased both sides and displeased no one with this head-in-the-sands policy—when, in truth, we have earned the suspicions of all.

It was not that Americans were opposed to change, but the U.S. government steadfastly asserted that it must come about through peaceful, evolutionary processes. American officials also argued that a colonial power could not continue to be held legally and politically responsible for a territory and yet have others dictating developments or controlling the pace of change in ways that might be detrimental to a peaceful outcome. Even if understanding were to overcome suspicion, neither of these high-principled and logical concerns could have much impact on the feelings of impatient nationalists. The American argument came across as a legalistic rationalization of the status quo.

Moreover, as is often the case in complex relationships, with colonies as with children, those in charge tended to underestimate the ability of those who were learning to manage their own affairs. It is easy to undervalue the contribution that experience can make to knowledge. The issue was complicated for the United States when the alleged unpreparedness of colonial people and other pretexts were used by Portugal and France to resist any change at all, and when Britain used the special status of the Central African Federation to block progress toward self-determination in Southern Rhodesia. Largely by condoning or overlooking the acts of others, the United States came to be identified with the forces opposed to nationalist aspirations.

This alienation of America from the decolonization process was exacerbated by ideological trends among the nationalists. Most states made the transition to independence peacefully and without undue tension. However, in those situations where the struggle turned to violence, the process was inevitably radicalized. In Algeria, the Portuguese territories, South-West Africa (Namibia), and, to some extent, Southern Rhodesia and Indochina, such was the resistance to self-determination that nationalist movements turned to or accepted communist offers of support, frequently after having made unsuccessful appeals for arms from Western countries. Some of those fighting for independence held Marxist views from the start, of course, given their experience with colonial exploitation, but even many of the moderate nationalist leaders were radicalized by their difficulties and by their disillusionment with what they interpreted as America's unwillingness to exercise its power to force change. For certain U.S. observers, there was an element of self-fulfilling prophecy in all this: Third World radicalism became a cause, as well as a result, of American hostility.

The resort to arms by nationalist forces to attain their goals did not usually sit well with the United States, even further removed from its own revolutionary beginnings, saddled with the status quo attitudes of middle age, and torn between loyalty to principle and allies. The flirtation with the Soviet Union in search of arms might have been understandable—though note Kennedy's warning in his Inaugural Address, that "those who foolishly sought power by riding the back of the tiger ended up inside"—but American officials and politicians could not understand how nationalist leaders allowed themselves to parrot the rhetoric of the Communists or to stand neutral on issues where the Communists were obviously at fault. The United States was aghast at the determination of some to reject entirely Western values, economics, and political systems just because of bad experiences with colonialism in the past. Similarly, any talk of renouncing debts owed to Western governments and banks or nationalizing foreign investment was unacceptable to America, however unjust the debt or exploitative the investment might have been. The nationalists were expected to forgive, forget, and remain open and loyal to the West.

Many of these currents came together in the movement known as neutralism (later renamed nonalignment), which grew out of an international conference of twenty-nine emerging nations of Asia and Africa convened in Bandung, Indonesia, in 1955, largely at the instigation of Gamal Abdel Nasser of Egypt, Jawaharlal Nehru of India, and Josip Broz (Tito) of Yugoslavia. The spirit of Bandung was symbolized by an expressed determination not to choose sides in the great global power struggle that had evolved after World War II. But the prominent role played at Bandung by Communist China made many Americans skeptical, and some people in powerful positions, including Secretary of State John Foster Dulles, simply could not accept the premise of Bandung; they concluded, in effect, that "if they [the so-called neutral nations] are not for us, then they must be against us."

Even a quarter-century of experience after Bandung did not seem to help Americans develop a durable, sophisticated understanding of nationalism. Whenever a decolonization struggle deteriorated into conflict, the United States found itself in an ambiguous, uncomfortable position, unable to understand the causes of violence. Officials talked a good (and familiar) line about supporting self-determination and majority rule in southern Africa, but when a Marxist faction won the civil war in the former Portuguese colony of Angola, for example, Washington seemingly could not accept the reality, so it would not recognize the regime. In Namibia, the Reagan administration could not cope with the prospect that the Marxist-oriented South West Africa People's Organization (SWAPO) was the leading candidate for power, and was even willing to oppose SWAPO in a way that risked creating the impression that the United States favored some form of continued control over Namibia by the white-minority regime in South Africa. Earlier there had been influential support in America for continued white rule in Rhodesia or for a government acceptable to the white minority; the will of the majority somehow ranked second as a consideration after the perceived alignment of key nationalist leaders with the Soviet Union or other communist powers. The same blinders were evident when, during the 1984 presidential debates, President Reagan described the opposition to Philippine President Ferdinand Marcos as communist inspired; he made no mention of the dictatorial and corrupt practices of the Marcos government.

Expressions of nationalist anger did not arise only in the midst of decolonization. They also arose where the European colonial power had long since departed, as in Libya; where a repressive, even feudal traditional leader was overthrown, as in Ethiopia; or where an entrenched dictator was ousted, as in Cuba, Iran, Nicaragua, and the Dominican Republic. We shall devote greater attention to Cuba below. At this point it is simply worth noting that in each of these cases the United States was particularly close to the ousted government—well aware of, and sometimes clearly embarrassed by, the slow pace of change and the failure of the regime to respond in a timely manner to legitimate criticism. Many Americans welcomed the changes, even if they came about through undemocratic means. Few would argue that the violence that occurred along the way or resulted later was necessary, or could defend the failure of the nationalists in all but the Dominican Republic to bring the people into the decision-making process through some form of democratic elections. Certainly it was difficult for Americans to understand the radical nature of some of the new governments, their close collaboration with the Soviet Union, and their adoption of communist ideology.

America's relations with the newly independent, highly nationalistic countries were bound to be difficult. They had little memory of the favors done for them yesterday, and they could not comprehend the twists and turns of American policy or the American obsession with the threat of communism. At the same time, an influential American perception of the newest countries and of political developments in other fiercely nationalistic states was negative: Their leaders were self-righteous; they squandered their resources and mimicked the life-styles of the people they had replaced; they failed to follow democratic traditions (Africans had "one man one vote one time," according to a popular slogan); they adopted socialist practices that were inappropriate for their societies and excluded Western capital; they were too frequently willing to ally with anyone—often the Soviets—no matter how ruthless, if only their own narrow interests and short-term objectives were satisfied; and they refused to play the game according to the rules of a gentleman's club or the Council on Foreign Relations. Their rhetoric was sharp, harsh, frequently bitter, and showed little respect for historical fact or Western concepts of fair play.

Many of these impressions were unfair, but some were not. In the euphoria of independence, abuses were too easily excused or explained away, and that sometimes discredited the entire decolonization process for people with short memories of their own country's problems. Americans, who increasingly got their news from brief and graphic television coverage, or from urgent stories by print reporters fresh off the plane with little cultural or political perspective, came to see nationalism negatively. The rise of corruption in the new nations and the emergence of repressive, undemocratic practices were a further blemish. Critics were jailed, the press was restricted, one-party governments were entrenched, and in many places elections were postponed indefinitely, if not formally eliminated. American and other Western disenchantment was nourished further by disgruntled nationalists who had their own grievances concerning developments in their countries. Thus, in the 1960s, the people who wanted to split off Katanga (later Shaba) from the Congo (later Zaire) emphasized their anti-communist, procapitalist leanings. The rebels who fought to convert the eastern part of Nigeria into "Biafra" did the same, as did the National Union for the Total Independence of Angola (UNITA), led by Jonas Savimbi, in the 1970s and 1980s.

Few situations rankled the American government and public as much as that of Cuba, only ninety miles from Florida, after the overthrow of dictator Fulgencio Batista and the rise of nationalist Fidel Castro. As noted earlier, the first American intervention in Cuba took place in 1898, and the hand of Uncle Sam was never far away for the next sixty years. The influence of American businessmen with interests in Cuba and of Cubans living in the United States was always strong, and the influence of conservative Cuban-Americans on U.S. policy in the hemisphere remains strong today. Few people now would defend the corrupt and ruthless practices of Batista, who assumed power in a bloodless coup in 1952; but it must be remembered that the United States was closely identified with Batista and, as has often been the case, began to withdraw support from him only after it was too late to have an effect. When Batista fled the country on New Year's Day of 1959, he had already looted its treasury. The coalition of forces that succeeded him, led by Castro, was bound to lash out against the United States.

Thus began a chain of events that was to have profound and far-reaching effects. Castro's coalition broke down and, as would happen later in Nicaragua and Iran, those who had been united against the common enemy soon became bitter enemies themselves. The new opposition to Castro carried on its fight from exile in the United States. It found support among those who objected to the public trials and executions of Batista supporters and, even more so, among Cuban and American businessmen whose land and businesses were expropriated under agrarian reform and other, draconian measures. The situation became mutually reinforcing; the United States was bound to lash out against Castro.

Castro may have begun as an idealistic nationalist setting out to overthrow a rightist government and help his people attain more economic and political control over their own country, but he ended up doing much more than that. Although Philip Bonsal, former U.S. ambassador to Cuba, has written that the Cuban Communist Party was a late supporter of Castro[7] (who later said that he had been a Communist all along), it was Castro who actually introduced the Soviet Union to the Western Hemisphere. He became an active supporter of revolutionary change elsewhere in the Americas, a Soviet surrogate in tense places in Africa, and the most vocal spokesman for Soviet policy within the nonaligned movement. It still riles Americans that such a minor offshore nation can remain so immune to the exercise of American power and can continue to cause mischief in Latin America and far-off Africa. For some American idealists, it is disappointing that Castro himself could betray his own revolution; for U.S. officialdom, it was especially aggravating that he could create a perceived threat to American security through his alliance with the Soviet Union.

Inevitably, the hemisphere which the United States once dominated, with much grumbling but little dissent, has become more divided. To be sure, rightist regimes can no longer count on automatic American support, but neither can the United States rely on anything approaching unanimous Latin American backing for its policy initiatives. Castro's early actions—including the acceptance of Soviet missiles—nearly led to a nuclear war, but that episode was short-lived. In other important respects events in Cuba have profoundly influenced U.S. attitudes toward nationalism and radicalism in the hemisphere. Each American president since Eisenhower has been intent on assuring that "another Cuba" does not occur on his watch. There has been a closing of the hemisphere's political space, with the United States less tolerant of revolutionary movements and elected radical governments. American officials worked to overthrow the elected Marxist government of Chile; ten years later Ronald Reagan invaded Grenada to save it from Marxism, and he openly supports the overthrow of the leftist Sandinista government of Nicaragua. In this area, the United States seems uniquely willing to flout established principles of international law in order to halt anything that can be plausibly labeled a communist advance. Damaging the American image—or, for that matter, international institutions such as the World Court—came to have less importance than stopping the establishment of "another Cuba."

The Cuban issue also had an important effect on the American domestic political scene. The United States now has a large, influential, and politically conservative Cuban-American community; they have been joined by exiles and refugees from other Central and South American countries who are willing to provide political and financial support to help preserve rightist regimes and oppose the Sandinistas and others on the left. They have, in effect, joined many Americans of East European descent in an ad hoc anticommunist coalition. If that coalition had its way, the United States would have little to do with communist-ruled states anywhere.

In the United Nations, the impact of decolonization was particularly noticeable. The organization evolved from a small, relatively homogeneous group of established nations easily dominated by the United States to an institution three times its original size and rather heterogeneous in composition. The new nations, dominant in number but not in financial or political power, had their own concepts of and priorities for the United Nations. "Legal niceties" was how the majority of new members, supported by many communist countries, described rules such as the requirement for a two-thirds majority to commit the organization on one or another side of an "important question." Once again the United States found its sincerity doubted when it tried to honor human rights at the same time it respected traditional rules it had helped to write. Thus, when South Africa's credentials were called into question at the United Nations in 1973, the United States argued that the nature of the South African government's racial policies had nothing to do with the legal status of the government issuing credentials. The next year, when South Africa was actually suspended from the General Assembly, U.S. Ambassador John Scali argued that this action would set a "shattering precedent which could gravely damage the U.N. structure." In the Third World's view, the United States was still arguing legalisms when more important and overriding principles, such as racial equality, were the real issues at stake.

That the newly independent countries and those who had overthrown governments perceived as Western dominated would have much in common was obvious and predictable. So was the natural antagonism that both groups felt toward their former oppressors. Anxious to assert their total independence from the colonial powers and the new leader of the West, the United States, they had to establish their own patterns of conduct; at the same time, their anger was ripe for exploitation. The Soviets were quick to pick up on the grievances of the newly independent countries against the West. Moscow provided little in the way of economic assistance and only just enough arms to insinuate itself in the Third World. But the Soviets delivered large quantities of rhetorical and political support. There were plenty of traditional, and especially regional, feuds in which the Soviets could intervene on one side or the other, such as the disputes between India and Pakistan; among Somalia, Ethiopia, and Kenya; between Morocco and the Polisario in the Western Sahara; and between Israel and the Palestinians, to name a few.

This Soviet opportunism compounded American unease over the years about the neutralist movement. For U.S. officials trying to cope with a world that seemed ever more unmanageable, the crucial question to other nations was, "Are you with us or against us?" Instead of answering unequivocally, the new countries and other leading "neutrals" tried to hedge and carve out an independent course. Admittedly, their motives were sometimes mixed. Yugoslavia sought to escape the smothering embrace of the Soviet Union. India, the world's largest democracy, had enormous problems and wanted to receive assistance from both sides—ultimately leaning heavily on the Soviets because of American support for Pakistan. Egypt, for its part, wished to lead the Arab world.

For John Foster Dulles, who dominated American foreign policy from 1953 until his death in 1959, the key period for the emergence of neutralism, the whole concept was immoral. In a June 1956 speech at Iowa State College, Dulles argued that widespread U.S. security arrangements "abolish, as between the parties, the principle of neutrality, which pretends that a nation can best gain safety for itself by being indifferent to the fate of others. This has increasingly become an obsolete conception, and, except under very exceptional circumstances, it is an immoral and shortsighted conception."[8] His world was bipolar, not multipolar, however much one might wish otherwise. For Dulles, there could be no middle ground between East and West, good and evil, right and wrong. How could one be neutral toward the Soviet Union while it was engaged in outrages in East Germany, Hungary, Poland, and Czechoslovakia? Dulles's successors and spiritual heirs were similarly shocked in later years when some members of the nonaligned movement seemed ready to condone the Soviet invasion of Afghanistan in 1979 or the Soviet-inspired suppression of the new trade union movement in Poland during the early 1980s.

In reality, of course, there has always been, and probably always will be, a debate among the nonaligned about their political attitudes toward East and West. The debate started at the Bandung conference and continues today. The nonaligned tend to be united by the attributes that make them a natural group: their recent attainment of independence, their underdevelopment, and their relative powerlessness as individual nations. On specific issues they tend to give wide berth, at least rhetorically, to the concerns of regional subgroups among them—to the Arabs on the Middle East, to the Africans on southern Africa, and to the members of the Association of Southeast Asian Nations (ASEAN) on topics that concern them. Efforts such as those by Cuba in 1979 to find and reinforce some "natural alliance" between the nonaligned and the Soviet Union have generally met with concerted opposition within the group. Indeed, most members of the nonaligned movement sharply criticized the Soviet Union for invading Afghanistan (over the objections of Cuba, then temporary leader of the movement), because they believed that the Kremlin had violated the independent status of one of their own brethren. (Their inability actually to do anything about it, beyond issuing declarations of outrage or disappointment, was another matter; they learned that the Soviet Union did not especially care about their opinion.)

Few Americans know of the inner workings of the nonaligned movement. But even those who do find it hard to explain how allegedly neutral nations could select Castro as their spokesman (or, for that matter, how the members of the Organization of African Unity could pick an Idi Amin or a Muammar al-Qaddafi as their chairman, even for a one-year rotation). Why is it, knowledgeable observers ask, that the nonaligned do not follow their own advice to the United States? Why do they not choose principle over procedure? How can they ignore the outside world's perception of them?

To return to a basic question at the heart of American foreign policy today, how did the United States change from a staunch supporter of nationalism and self-determination to a country with strained relations with developing nations, some of whom would not even credit the earlier American position? One thing that happened is that the United States found itself unable to implement its own ideals; despite its considerable political, military, and economic power, it was not in a position to right old wrongs or unilaterally to change prevailing international circumstances. America simply could not fulfill the ambitious expectations of the nationalists. On the contrary, the United States ended up accommodating the views of the colonial powers to a degree that would have shocked Franklin Roosevelt, in part because of the emergence of the Cold War and the heightened international security responsibilities that were taken on by Washington in the postwar vacuum. Inevitably, American officials, obsessed with a fear of communism, hesitated to apply pressures that might weaken an ally or cause it to reduce its contribution to Western defense. Finally, the level of mutual tolerance and understanding between the developing countries and the United States was very limited.

Despite this background, the current American estrangement from the newly independent and assertive countries of the Third World is still little understood; few people appreciate its role in America's predicament in international affairs. It may be comforting to believe that the righteous policies of the United States are opposed only by an anti-American, pro-communist, and largely backward majority in the United Nations, but the reality is quite different. Today many of America's closest allies—former colonial powers of the first rank—have joined the developing countries in criticizing U.S. policy on relations with the Soviet Union in general and arms negotiations in particular, on the Middle East and Central America, and on the law of the sea and other matters of North-South conflict. Ironically, they have emerged from their experience as colonialists with a greater understanding of nationalism than the Americans who used to preach to them about it. They see the United States as too dogmatic, too likely to view developments through an East-West lens, and too prone to ethnocentrism. To accuse them of an opportunistic change of heart is too easy; it overlooks their own willingness to acknowledge and correct past mistakes.

Under the same circumstances, in pre–World War II days the United States might have returned to a traditional posture of isolationism. Now, instead of reexamining foreign policy, it has resorted to a new nationalism of its own. A tough, macho language characterizes the American discourse—even though, as in the case of the hostage crisis in Iran and the attacks on American marines in Lebanon, the rhetoric cannot be backed up with effective force. Reluctant to send troops off on new overseas adventures, the American public nonetheless supports a military buildup—as if that were the answer to U.S. political problems in the world. Congress sensibly hesitates to go along with overthrowing the government of Nicaragua, but nearly everyone seemed to support direct action to install an acceptable government on the tiny Caribbean island of Grenada, because it somehow made besieged and beset Americans feel better about their country's place in the world. One symptom of the new American attitude is a retreat from multilateralism, of which the United States was once a principal sponsor. Now America goes it alone, pushing for the acceptance of American ideas and institutions, rather than trying to persuade by example. There is in the United States a new nationalism which seems to require strident repetition of American values and views, to the point where Washington almost appears to be imitating Moscow. If truths that once were self-evident now have to be sold like used cars, there must be something very wrong.

5 Decade of Determination

J. BRYAN HEHIR

J. Bryan Hehir, secretary of the Department of Social Development and World Peace of the United States Catholic Conference, is a theologian and scholar whose writings on ethical questions in international relations have had considerable influence. From 1981 to 1983 he worked with the National Conference on Catholic Bishops' Ad Hoc Committee on War and Peace, drafting its pastoral letter on nuclear deterrence. A columnist for *Commonweal*, Father Hehir is also a senior research scholar with the Kennedy Institute of Ethics at Georgetown University and a research professor of ethics and international politics at Georgetown's School of Foreign Service.

THE 1980S IN THE United States, dominated as they are by the philosophy and the personality of Ronald Reagan, are often thought to be the antithesis of the previous two decades. The contrast in domestic social and economic policy may support this description, but the foreign policy agenda bears closer analysis. The political and moral staying power of the 1960s is evident in the way foreign policy questions continue to be defined, in the terms and categories used, and in the persistent influence of certain key policy decisions of that era.

The current debate over nuclear weapons, perhaps the central foreign issue facing the Reagan administration, has been cast in categories developed in the 1960s. One of the key arms control issues of that decade, defensive systems, has recently resurfaced. The NATO debate over the installation of cruise and Pershing missiles during the late 1970s and early 1980s sounded at times like a rerun of the 1960s, with Europeans worrying about U.S. nuclear guarantees being divorced (or "decoupled") from Europe and the United States pressing for an increased European conventional commitment. In parallel fashion, the Central America debate has been framed in terms of the Vietnam experience. Critics of U.S. policy raise the specter of another "quagmire" for the United States. Supporters of a tougher line in Central America urge that the United States be prepared to demonstrate that its "Vietnam syndrome" has been put to rest.

One of the most visible recent characteristics of this debate has been an explicit appeal to moral argument, particularly with respect to the nuclear question and the ethics of intervention. Here again the parallel with the 1960s is striking. If anything, ethical arguments are more common in the 1980s, but they draw heavily on the discussions of the 1960s.

The complex relationship between these two decades can be valuably illuminated by examining, in turn, the policy vision of the 1960s and the nature of the world it tried to influence, two major issues which engaged this policy vision, and the political and moral legacy of the 1960s, as expressed in the 1980s.

The place to begin to understand the foreign policy perspective of the 1960s is President Kennedy's Inaugural Address. It has come to be viewed as the classical expression of the sense of hope and expectation which the Kennedy team brought to office. In the post-Vietnam era, it has been criticized more than praised, but this view fails to do justice to the way the address captured the mind and spirit of the country in the winter of 1961.

Examining the Inaugural Address from the perspective of the 1980s, one is struck by the primacy given to foreign policy. The speech lacks any reference to economic policy, taxes, race relations, or any of the "social issues" so prevalent in American political debate today. Kennedy defined the goals of his presidency almost exclusively in terms of America's role in the world. That world, as Kennedy saw it, was both very different and more dangerous than the one confronted by previous generations of Americans, "for man holds in his mortal hands the power to abolish all forms of human poverty and all forms of human life." The centrality of foreign policy in the Inaugural reflected a widespread conviction in the early sixties that the United States was "called" to fulfill a unique role in the world, and an even deeper conviction that it could accomplish whatever it set its mind and its technology to do. The address was specifically directed to allies, adversaries, and "new states"; to each Kennedy pledged cooperation, and with each he proposed the initiation of a new relationship with the United States. The most remembered and most controversial passage defined America's commitment to "pay any price, bear any burden, meet any hardship, support any friend, oppose any foe to assure the survival and the success of liberty."

This passage graphically conveys the essential distinction between the 1960s and the 1980s. One can conceive of an American president, in a moment of rhetorical flourish, uttering these words today, but their impact both on the policy debate and on public opinion would differ significantly from their reception in the 1960s. The prevailing conception of the U.S. role in the world has become more modest, and that modesty is derived in part from the experience of the 1960s. A decade that began with an extraordinary commitment to foreign policy ended with the nation divided in an unprecedented manner over a foreign commitment. The foreign policy story of the 1960s is captured by the contrast between the soaring rhetoric and idealism of the Inaugural and the frustrating catalog of crises which spanned the decade, from the Bay of Pigs to the

Gulf of Tonkin to the Gulf of Aqaba. Those ten years saw some authentic achievements in American foreign policy, but the dominant note struck today centers on the lessons learned about the limits and liabilities of being a superpower.

The divergence between what was proposed and what came to pass in the 1960s can be attributed both to the state of the world and to the content of U.S. policy. What was the American design, and why was it not fulfilled?

Kennedy inherited the Cold War configuration of the 1950s, in which the two superpowers were locked in a global competition with no other countries politically or militarily prepared to challenge them. In their tone and their concerns, the early sixties were very much part of the Cold War. But even as Kennedy took office, forces in the international system were working to transform the bipolar competition. The Inaugural Address expressed a hope and a confidence in the ability of the major powers to shape the world in a positive direction, for the benefit of all. But the character of international politics was moving from the stark and dangerous clarity of the Cold War to a more complex, though no less dangerous, pattern of relations. The changes made it less likely that either superpower could unilaterally shape the world for good or for ill.

The world which the Kennedy administration sought to shape had three distinct levels of interaction.[1] The two superpowers were historically unique, and their relationship was the central factor in world politics. The middle powers were usually allies of one of the major powers, either by choice, coercion, or chance. They were restless in this role but lacked the means to raise their status. The small powers were spread throughout the international system, but their significance was greatest in the developing world, an entirely new arena for policy in the postcolonial period.

The essence of superpower status was the ability to act simultaneously on all three levels of the international system. Entering the 1960s, the superpower relationship remained the essentially bipolar pattern of competition which had emerged after World War II. Historically, such a configuration of power had bestowed on each major power the ability to influence the rest of the international system directly and decisively. Two crucial factors came to place restraints on the range and degree of superpower activity in the 1960s, however: the nature of nuclear weaponry and the growing pluralism of international political relations. The first sign of these transformations was the novel impact of nuclear forces on traditional notions of foreign policy. For centuries it had been presumed that military might granted political influence, and that increasing one's supply of power enhanced one's capacity to direct the course of world events. Nuclear weapons snapped this linkage between strategy and politics. Henry Kissinger later expressed the new conventional wisdom about war and politics:

> As recently as twenty-five years ago, it would have been inconceivable that a country could possess *too much* strength for effective political use. . . . The paradox of contemporary military strength is that a gargantuan increase in power has eroded its relationship to policy. The major nuclear powers are capable of devastating each other. But they have great difficulty translating this capability into policy except to prevent direct challenges to their own survival. . . .[2]

This passage indicates a truth about nuclear weapons that became much more evident in the 1960s than it had been during the previous decade: The very destructiveness of these new military instruments, their capacity to threaten the user as well as the adversary, rendered them irrelevant to many of the key foreign policy questions of the decade. Qualitative superiority in nuclear weaponry did not provide a license to use this awesome power effectively; the character of nuclear weapons made them unusable. No one doubted that being a nuclear superpower conveyed a capability to threaten the entire structure of international life, but no rational purpose could be served by executing that threat. The awareness of the risk of invoking the nuclear moment made the superpowers particularly cautious in situations of direct confrontation, such as the crises involving Berlin and Cuba. In these rare but frighteningly direct encounters, the stark reality of the nuclear danger suddenly became tangible for statesman and citizen alike. Confronted with this danger, both superpowers have assumed surprisingly conservative foreign postures. The paradox of the nuclear age has not been that these weapons eliminate international competition or even the incentive to take occasional risks to gain marginal advantages, but rather that nuclear weapons have put in place new and fragile limits on how that competition is pursued.

The development of nuclear weaponry had a second major impact on the role of the superpowers in world affairs during the 1960s, the rise of international political pluralism. The Cold War patterns of the previous two decades had left little room for allies and other middle powers to play a significant role in world affairs.

The scope of superpower action and the threat of superpower confrontation led most countries either to seek alliance with a major power or to avoid entirely being drawn into the competition. In the 1960s a new level of international activity became visible just below the superpower face-off. Some of the middle powers recognized both the paralysis of the superpower nuclear deadlock and the limited freedom this allowed for their own independent initiatives. There was no question of any nation challenging the military dominance of either nuclear giant; the opportunity of the 1960s was for the reassertion of international autonomy for countries that had been dependent or essentially dormant since World War II. It is true that the possibilities were much greater in the West than the East, but there was a certain parallelism between France's recalcitrance in the Western alliance and the Chinese challenge to the Soviets.

In the West, the postwar endeavors of the United States created the conditions for a revival of Western Europe and Japan, making them the premier middle powers of the world. America's position did not, however, facilitate its adjustment to France's independent conception of its strategic needs in the 1960s or Japan's assertive view of its role in the world economy of the 1970s. In both cases, revived middle powers challenged the nature of American leadership of the alliance. It was far easier to revive defeated allies than to relate to mature middle powers. The fact that the Soviets were even less successful with the Chinese was only small consolation.

It was de Gaulle who most firmly grasped the logic of the nuclear dilemma faced by the superpowers. The partial paralysis induced by the nuclear danger enabled him skillfully to carve out space for France to act independently in Europe without ever quite forsaking the protection of the American nuclear umbrella. Recognizing that the American commitment to Europe was a reality which the United States neither could nor would unilaterally abandon, de Gaulle did not feel obliged to follow the direction of Kennedy's "Grand Design" for a united Europe as the price of protection. De Gaulle's objections to the Grand Design were both political and strategic. Politically, he believed it eroded France's identity as an independent political actor in Europe and on the world stage. Strategically, he feared that the caution induced by the conditions of deterrence could work against the best interests of Europe at a moment of superpower crisis. This fear led de Gaulle to build a small nuclear force that could hypothetically serve to call into play the U.S. strategic deterrent in the event of a European crisis. This was a daring move which illustrated precisely how the bipolar nuclear standoff had spawned political pluralism. It prefigured in the 1960s the pattern of 1970s alliance pluralism, in which Japan and West Germany would challenge American economic supremacy.

The political pluralism embodied in French and Chinese dissent from superpower designs in the 1960s was reflected at a third level of the international system, among the developing countries. In the Third World, the constraints on superpower action were considerably less than in situations where a direct confrontation might result. Indeed, it was during the 1960s that the developing world became the field of play for indirect superpower competition. The disparity of power between the superpowers and their allies, clients, or adversaries in the developing world was enormous. But in this arena, as in Europe, power did not guarantee the compliance of other nations.

Two constraints limited the scope of superpower action in the Third World. The decolonization movement ruled out an imperialist style of policy.[3] The nationalist self-identity of newly independent states and the desire of both superpowers to be seen as supporters of Third World nationalism led each to tread lightly in its relations with the developing countries. Moreover, the enormity and complexity of the internal problems faced by the new states of Asia, Africa, and Latin America frustrated the implementation of superpower designs—from the most altruistic to the most self-interested. From the Congo to Cuba to Vietnam, U.S. policymakers found the developing world less fathomable and less pliable than a superpower of another age might have expected.

None of this is to say that U.S. or Soviet intervention could be forcibly prevented by Third World nations; superpower influence was pervasive throughout the sixties. Nonetheless, superpower influence could be wielded with even less certainty in the politics of the developing world than it could in the formal setting of the European alliance.

Returning to the Kennedy Inaugural Address, it becomes clear that, above all, Kennedy attempted to stress the responsibility of the superpowers to direct, control, and shape events. "Let both sides join in creating a new endeavor," he proposed, "not a new balance of power, but a new world of law, where the strong are just and the weak secure and the peace preserved." But the dynamic of change in international relations pointed less toward the exercise of control and the enforcement of coherence from the top and more toward complexity and obstruction imposed by newly emerging forces in the world.

Compounding the effects of these external forces, the American policies designed to implement the Inaugural vision had their own internal tensions. The Kennedy address was a clarion call for collective superpower action—"begin anew the quest for peace . . . formulate serious and precise proposals for the inspection and control of arms . . . invoke the wonders of science instead of its terrors"—and it pledged a U.S. policy committed to the defense of human rights, opposition to colonialism, and an end to the arms race. The address was a moral challenge to the nation, defining in a cool but strong moral tone the U.S. obligation to control arms and to help the developing world.

Coexisting with this moral commitment in Kennedy's policies, however, was an ideological edge and a determination to pursue peaceful objectives in an assertive manner. As the complexities of the nuclear standoff and the competition with the Soviets in the developing world became clear, the moral and the ideological were fused, emphasizing the confrontational dimension of the superpower relationship. The Inaugural's theme of a new beginning receded, and the pledge to "bear any burden . . . oppose any foe" in order to assure the survival of liberty came to the fore, leading to activist policies in situations which did not promise success.

It would have been difficult to fulfill the ambitious promises of the Inaugural Address even in a stable period of international relations. But in a setting which was revolutionary in character and nuclear in content, realizing the hopes of the Inaugural placed exceedingly high demands on American foreign policy. Moreover, that policy was premised on the conviction that sufficient energy and effort could be marshaled by the world's leading power to produce the desired results. The emphasis in the Inaugural, and in the policies which flowed from it, was on political will. The complexity of the international system and the obstacles others might pose to American progress were not highlighted.

It is not surprising, in retrospect, that the direction of events and the design of U.S. policy moved in ways the authors of the

Inaugural would have found disappointing and disconcerting. Two key policy dilemmas of the decade demonstrate the difficulty of realizing the promise of the original vision.

During the 1960s the United States was engaged in a bewildering array of issues ranging from the threat of war in the Caribbean to the fact of war in the Middle East and Southeast Asia, from the political problems of the Atlantic Alliance to the economic problems of the Alliance for Progress. Two issues concerning the use of force—the control of nuclear weapons and the politics of intervention—involved political and moral questions that pervaded American foreign policy during the 1960s and have emerged again in the 1980s. Both issues shed light on the policy process in the 1960s and both profoundly shaped America's foreign relations during the following two decades.

The threat of nuclear war was used more than once by Kennedy in his Inaugural Address to symbolize the qualitatively new challenges faced by American foreign policy. The issue of nuclear arms control remained for him a central concern. Two recent historical studies of the nuclear era agree in their assessment of the fundamental contribution that Kennedy policies made to the politics and strategy of nuclear diplomacy. In Michael Mandelbaum's judgment, "by circumstance and by temperament it fell to this administration to make the definitive American response to the problem of nuclear weapons."[4]

The response of the Kennedy administration was distinguished by the adoption of sophisticated concepts in the design of nuclear strategy. As Lawrence Freedman points out in his detailed study *The Evolution of Nuclear Strategy*, the United States has inherited much from the nuclear policies of the 1960s.[5] These include terms which still define the nature of the debate in the 1980s: *flexible response, secure second strike, mutual assured destruction*. The central figure in the development of this policy was Robert S. McNamara, secretary of defense for seven years under Presidents Kennedy and Johnson. In Freedman's view, "no single public figure influenced the way we think about nuclear weapons quite as much as . . . McNamara."[6] He drew extensively on a round of advanced research in nuclear strategy and diplomacy which had been under way since the late 1950s, explicitly shaping U.S. policy in terms of some of these basic concepts in speeches and congressional testimony.

The impact of the Kennedy-McNamara policy can be better appreciated if it is placed in historical perspective. Kennedy had campaigned against the Eisenhower legacy in nuclear policy, claiming that America was on the short end of a "missile gap" and that it was bound to the dangerous and paralyzing policy of Massive Retaliation (the policy of the United States that it was entitled to use nuclear forces at times and places of its own choosing). Ironically, the notion of the missile gap was revised soon after Kennedy took office, and a determination to reshape what remained of the Massive Retaliation strategy set the theme for McNamara's early years at the Defense Department. McNamara's impact was at once intellectual, organizational, and political.

The fundamental contribution of the early Kennedy period rested not in a single policy or program, but in a shift in the way nuclear weapons were understood and nuclear strategy defined. By the early sixties the nuclear debate was still in rudimentary form, though almost two decades old. Underlying the more routine issues of weapons systems, technology, and tactics was the essential question of how nuclear weapons affected the foundations of foreign policy.

Two broad responses to this question had emerged. One position defined nuclear weapons as a qualitatively new event in political and military history and argued for a recasting of the classical understanding of war as an extension of politics. The classical view, most cogently formulated by Karl von Clausewitz, a nineteenth-century Prussian officer, held that force could be used in a controlled fashion to achieve rational and morally justifiable goals. As a qualitatively different category of weapon, nuclear arms challenged the classical view. The most authoritative and eloquent challenge came from a careful student of Clausewitz, Bernard Brodie. In one of the path-breaking articles of the nuclear age, Brodie asserted: "Thus far the chief purpose of our military establishment has been to win wars. From now on its chief purpose must be to avert them. It can have almost no other useful purpose."[7]

The alternative view did not deny the radically different characteristics of nuclear weapons, but argued for continuity with early periods of policymaking. A consequence of this position was to retain the idea that nuclear weapons were usable military or diplomatic instruments.

These two broad responses to the nuclear phenomenon were

debated in strategic literature and in the U.S. government throughout the 1960s. It seems clear in retrospect that McNamara was philosophically sympathetic to the Brodie school. (In the evolution of the Kennedy and Johnson presidencies there was evidence that this affinity to the Brodie position did not always carry through in every policy decision. For example, McNamara was attracted, for strategic and moral reasons, to the counterforce position in 1962. He sponsored extensive studies on fighting and controlling nuclear war, and he announced—with a notable lack of enthusiasm—a ballistic missile defense program in 1967. However, these measures did not represent the dominant ideas in the McNamara years at the Pentagon.) Kennedy and McNamara made an early decision to reduce U.S. reliance on nuclear weapons and to define the purpose of the nuclear arsenal in limited and specific terms. Reducing American reliance on nuclear weapons did not necessitate a reduction of the size of the arsenal. Indeed, Kennedy presided over a major increase in U.S. nuclear striking power. Reduced reliance had consequences instead for the role of nuclear weapons in U.S. defense policy. The Kennedy-McNamara objective was to change the public impression, created by Massive Retaliation, that the United States was prepared to use nuclear weapons in a wide variety of circumstances.

In pursuit of the dual goals of limiting the strategic role of nuclear weapons and gaining more secure political control over them, McNamara constructed a policy around three ideas. First, the purpose of nuclear weapons was to deter an attack on the United States and its NATO allies. Deterrence was best served "by maintaining a highly reliable ability to inflict unacceptable damage upon any single aggressor or combination of aggressors . . . even after absorbing a surprise first strike."[8] This definition of the role of U.S. strategic nuclear forces gave them a critical but very specific function: It required a deterrent force that was geographically dispersed, protected in hardened silos, and securely under control.

The second element of the McNamara strategy was to expand the U.S. and NATO capacity for nonnuclear deterrence. This was necessary to avoid early or accidental use of nuclear weapons and to provide a more credible deterrent against nonnuclear threats. Known as the Flexible Response strategy, this plan for expanding conventional forces became one of the most controversial aspects of McNamara's policies. Many Europeans feared that it gave the wrong signal to Moscow and that it might foreshadow a U.S. intention to "decouple" the U.S. land-, air-, and sea-based deterrent from the European theater. McNamara pressed the issue within the U.S. government and NATO, achieving an agreement in principle in 1967.

The third characteristic of the strategy was related to the Flexible Response concept. It involved an effort to erect a graduated series of decision barriers between the use of conventional and nuclear weapons. These "firebreaks" were designed strictly to delimit the use of nuclear weapons by forestalling their early or accidental use. Both Flexible Response and the firebreak concepts made eminent sense in light of Kennedy's intention to reduce the significance of nuclear weapons. Yet both were used to some effect by General de Gaulle to argue the case in Europe that the superpower nuclear stalemate had made the American guarantee of strategic nuclear protection unreliable.

In the 1950s Henry Kissinger had lamented the lack of "doctrine" in American strategic planning.[9] By the mid-1960s McNamara, supported by both Kennedy and Johnson, had articulated a sophisticated philosophy for the foreign policy role of nuclear weapons. The McNamara strategy required extensive changes in the structure of America's armed forces, U.S. strategic doctrine, and categories of defense spending. His strategy encountered opposition in NATO, in Congress, and in the military; he did not get everything he wanted. But the changes wrought in American defense policy in the Kennedy-McNamara period were such that, in Mandelbaum's judgment, "the next thirteen years were . . . a series of footnotes to the Kennedy administration."[10]

The conviction that a survivable second-strike deterrent supplemented by a wider range of conventional options would provide the best possibility for preventing nuclear war in the event of a crisis was put to a major test in the Cuban missile crisis. The Soviet move to install intermediate-range nuclear missiles in Cuba touched off the most dangerous superpower confrontation of the nuclear age. The possibility of superpower war, submerged on a daily basis by mutual caution and the distractions of other relationships, became dramatically evident. The facts of the case—the Soviet plan to deploy the missiles, the U.S. discovery of the plan before it could be implemented, Kennedy's ultimatum to Khrushchev, and the cosmic drama of the thirteen days that ensued—are available in a still-growing body of literature.[11] It is important here simply to highlight two consequences of the crisis for American foreign policy in the 1960s.

First, the U.S. reaction to the missile threat, directed by President Kennedy with the aid of a select group of advisers, appeared to be a textbook illustration of how Flexible Response was designed to function. While analysts still dispute the reasons for the Soviet retreat, the effect of the incident within the Kennedy administration was nonetheless to increase confidence that force—or its threat—could be carefully applied in an incremental fashion to accomplish specific objectives.[12] The power of this conviction among key personalities on the Kennedy team subsequently affected policies outside the nuclear arena. Second, the experience for both superpowers of going to the brink of nuclear war also appeared to impel them toward new efforts to control the arms race. In a crucial exchange of letters, both Kennedy and Khrushchev used the crisis to call for another attempt to reduce the danger of nuclear war.

These pledges were not forgotten in the aftermath of the crisis. The influence of Kennedy-McNamara policies in the nuclear age was also felt in the shaping of a new arms control process. The arms control philosophy of the 1960s was at once more modest, more complex, and more sophisticated than its predecessors. Negotiations were normally carried on directly between the superpowers rather than under the United Nations aegis, as in the 1940s and 1950s. The political premise of these negotiations was that the nuclear stalemate had created a narrowly defined area of common interest between the superpowers. Raymond Aron referred to them as "les grands frères."[13] On the basis of this premise, specific objectives of mutual interest were to be defined and agreements achieved in a step-by-step manner. This carefully designed process yielded three favorable results in the 1960s: the Limited Test Ban Treaty (1963), the Non-proliferation Treaty (1968), and the beginning of the SALT negotiations.

In the United States, the intellectual foundation for this style of negotiations had been developing for some time among the same group of scholars and analysts who had provided the concepts McNamara used in reshaping the nuclear deterrent. Notwithstanding its considerable achievements, the arms control process might have had a far stronger impact on superpower relations and the nuclear arms race if U.S. foreign policy had not been substantially redirected in the middle of the decade to a totally different arena—a war of intervention in Southeast Asia.

The Vietnam challenge to American foreign policy spanned the three presidencies of the sixties. John Kennedy made the commitment, Lyndon Johnson expanded and escalated it, and Richard Nixon continued the policy into the 1970s. In the 1980s the Vietnam debate is still with us, as the literature on the lessons of the conflict continues to expand.[14]

To examine Vietnam as an illustrative case of U.S. policymaking in the 1960s requires that one address two key questions: America's definition of its purposes in Vietnam and its use of power in the war. The question of U.S. purpose engages the issue of how the United States evaluated the meaning of Vietnam in the context of world politics. Describing his own perspective as he assumed responsibility for Vietnam policy at the end of the decade, Henry Kissinger wrote: "In my view, Vietnam was not the cause of our difficulties but a symptom. We were in a period of painful adjustment to a profound transformation in global politics; we were being forced to come to grips with the tension between our history and our new necessities."[15]

This profound transformation in world politics could be more clearly assessed at the end of the decade than at the beginning. The transformation, as we have already seen, involved both an unprecedented qualitative increase in the military might of the superpowers and a decline in the capability of any single nation to shape the world according to its own particular conception of political order. The Kennedy Inaugural clearly signified an awareness of revolutionary changes occurring in the structure of world affairs. For both the Kennedy and Johnson administrations, the experience of Vietnam proved to be a searing political and moral demonstration of how revolutionary change could frustrate even the most powerful nation.

The use of military power in Vietnam demonstrated the complexity of the policy challenge. One revisionist interpreter of the Cuban missile crisis contends that the way U.S. success was interpreted in that event "contributed to President Johnson's decision to use American air power against Hanoi in 1965."[16] This parallelism may be a bit simple, but it does point to an interesting paradox of U.S. policy. The very success which the Kennedy administration and many analysts saw in the strategy of the missile crisis bred a confidence in the ability to use the same approach in other situations. The concept of employing graduated, escalating pressures, carefully correlated with political proposals for negotiation, had apparently achieved substantial results in a superpower setting. The missile crisis became a model of the conflict resolution process. The application of this model to the less predictable course of a social revolution in a developing country proved more difficult than anyone had imagined.

In exploring the purposes of American intervention in Vietnam, it is hard to overestimate the influence of the American view that its role in the world would be significantly affected by the outcome. Neil Sheehan's commentary on the Pentagon Papers identifies this persistent theme of the policy process. In that Defense Department study of U.S. policy in Vietnam, Sheehan noted, "the war was . . . considered less important for what it meant to the South Vietnamese people than for what it meant to the position of the United States in the world."[17]

The way in which policymakers related their decisions about Vietnam to the U.S. role in the international system changed during the course of the war, but the differences were always variations on a constant theme. There were basically two versions of the argument that Vietnam was crucial to U.S. policy worldwide. They were distinguished by the identity of the adversary. Early in the war attention was focused on Soviet intentions, but by the mid-1960s the principal concern shifted to the People's Republic of China.

In 1961 Vietnam was at best a dependent variable in the American foreign policy equation. The issue was not near the center of foreign policy concerns for the new Kennedy administration, and early decisions on Vietnam were in great measure determined by events in other parts of the world. The dominant theme of the foreign policy agenda was the intense and dangerous superpower competition, exacerbated by a bellicose Khrushchev intent on testing the inexperienced Kennedy's mettle.

The focal point in 1961 was the August–October war of nerves over Soviet moves threatening the Western position in Berlin. The escalation of tension was vividly symbolized by the Communists' construction of a wall dividing the city into Eastern and Western sectors. Prior to the Berlin crisis three events in the spring of 1961 began to set the stage for Vietnam decisionmaking. The first, a persistent problem in 1961–62, was the deteriorating situation in Laos, where a Western-backed faction was losing ground against the communist-supported Pathet Lao. The second was the failure of a U.S.-sponsored invasion at Cuba's Bay of Pigs in April 1961. The third was the June meeting of Kennedy and Khrushchev in Vienna, in which the Soviet leader threatened "to solve" the Berlin problem by signing a World War II peace treaty with East Germany. The cumulative impact of these events was to create a sense of insecurity among American policymakers, derived from the fear that Khrushchev might underestimate the determination and courage of Kennedy or the American people. Consequently, there arose a conviction in Washington that the United States had to take a stand somewhere in the world to demonstrate its willingness and ability to resist Soviet intimidation.

The need to confront the Soviets in a convincing manner was perceived as a twofold challenge. First, the United States had to resist Khrushchev's moves at the very center of superpower competition, in Berlin. Second, Soviet designs had to be opposed somewhere in the developing world. Kennedy felt a special urgency concerning this second test of will because of the speech Khrushchev had given on January 6, 1961, in which he proclaimed nuclear and conventional wars to be irrational, but espoused "wars of liberation" as part of the Soviet global revolutionary design. Several commentators have pointed to the impact of this speech on Kennedy.[18] He was convinced of the inevitability of rapid social change in the developing world, and he saw it as a testing ground for the competition between the Soviet and American systems.

However, the Soviet view of wars of liberation transformed the expected ideological competition into a strategic test of strength. Kennedy had campaigned against the Eisenhower-Dulles doctrine of Massive Retaliation because it was vulnerable to "salami style" tactics of aggression. Small but important gains by adversaries would always seem disproportionate to invoking the nuclear threat and would thus go unopposed. Especially after Khrushchev's speech, Kennedy was determined that the Soviets not find a means of using just such tactics against his policies. He had expanded the counterinsurgency capabilities of the United States early in his presidency. He now seemed predisposed to choose a time and place where he could use these capabilities to demonstrate to the Soviets that they would not have a free hand in the developing world.

The deteriorating position of President Ngo Dinh Diem in South Vietnam brought several of these issues to the fore. The need to confront the Soviets, the need to prove America's fidelity to its commitments, and the need to fashion U.S. strategy to meet unconventional warfare all made Vietnam appear to be a compelling test case. This was the rationale that generated Kennedy's 1961 and 1963 commitments of political support, military assistance, and U.S. advisers for the Diem government.

By 1965 Vietnam had become a test of U.S. willingness to counter the even more aggressive Chinese view of wars of liberation. In September Marshall Lin Piao, the Chinese minister of defense and deputy premier, gave a speech in which he divided the world into the rural proletariat and the urban affluent, who were ultimately to be encircled and overwhelmed by the poor. This speech from a major Chinese figure had an impact on American officials similar to that of the Khrushchev speech on Kennedy in 1961. Faced with this global revolutionary strategy, in which Vietnam was perceived to be the first step, American policymakers constructed an equally cosmic rationale for U.S. involvement in Vietnam: An American presence was necessary there to defend South Vietnam, to protect other Asian nations, and to prevent the spread of subversion to other small and weak nations in the developing world.

This threefold rationale was reinforced with two historical analogies, Munich and Korea. Munich demonstrated the hazards of failing to resist aggression, and Korea illustrated what could be accomplished through determined resistance. These analogies were used throughout the 1960s by President Johnson and Secretary of State Dean Rusk. Both were convinced that a failure to draw a clear line in Southeast Asia would foster global instability and would lead to the questioning of America's will by its staunchest allies. In light of this conviction and the "lessons" of the 1930s and Korea, the United States committed more than four hundred thousand troops to Vietnam by the mid-1960s. The irony of this policy was that it led precisely to the questioning of American military capabilities and political wisdom which it was intended to avert.

The rationale for U.S. involvement was clear, but a rationale is not a policy. The policy problem in Vietnam was therefore not why the United States was there but whether its objectives were attainable. The essence of the policy debate was the relationship of ends and means. Stanley Hoffmann captured the core of the problem in 1969, when he wrote, "We have fought a war for objectives that were unreachable. . . . The more those objectives eluded us the more we escalated our means, without realizing that the means we used made our goals even more unreachable and destroyed any chance there might have been of getting near them."[19]

The gap between ends and means was rooted in three factors. The South Vietnamese government could not earn any legitimacy among its populace, and it was never able to conduct an effective war on a national basis. One did not then—and does not now—need to approve of the harsh authoritarian quality of the North Vietnamese government to recognize that its conduct of the war had a discipline and a determination not matched in the South. Psychologically, the North Vietnamese were committed to the war without reservation; it was for them a continuation of decades of struggle with foreign powers—the Japanese and the French before the Americans. The United States could never match the intensity of this commitment, for its stakes were never the same. Militarily, U.S. expertise and weapons technology could not overcome these political and psychological disadvantages, and as a result there was a continuing need for more men and firepower.

The bombing policy debate uniquely illustrates the ends-means dilemma the United States faced in Vietnam. More powerfully than any other policy question, it raised the issue of how the United States conceived the relationship between its political purposes and its use of power. The air war began in Vietnam in February 1965; it continued, with pauses and interruptions, until the end of the war and proved to be the most controversial dimension of U.S. policy.[20] The political and moral debate over bombing policy engaged both policymakers and the public. The key figure in the debate was Secretary of Defense McNamara, who first directed and defended the policy and then led the fight within the government to change it. The premises of McNamara's strategy were shaped by his experience rationalizing and controlling the nature of the American nuclear arsenal in the competition with the Soviets. David Halberstam describes how the McNamara team perceived its task in Vietnam:

> They would, they thought, use power in the same slow cautious judicious manner. Not too much, not too little—signaling their intentions, that is, that they did not want to go to war; rejecting the radicals on both sides; being in control of the communications all the way through.[21]

The title given to the bombing strategy, "The Progressive Squeeze," conveyed its rationale. The bombing of North Vietnamese targets was intended to supplement the war in the South, not substitute for it. It was designed to accomplish a limited objective and was supposed to be limited in its scope. The latter intention was not sustained, partly because of the unexpected resistance of the North Vietnamese and partly because there had never been a consensus within the Johnson administration on the purposes of the bombing.

The Joint Chiefs of Staff did not share McNamara's view of the role of air power. At various times they advocated shifting from his incremental strategy to a "sharp knock" strategy designed to break the will of the leadership in North Vietnam. The practical differences between the two policies are illustrated by contrasting the pre-1968 bombing program with the Christmastime bombing of Hanoi and Haiphong in 1972. The latter involved the kind of air strikes the military had lobbied for much earlier in the war. Admiral Thomas Moorer, chairman of the Joint Chiefs of Staff, convinced that the Christmas bombing brought Hanoi to the negotiating table in 1972, declared: "Airpower, given its day in court after almost a decade of frustration, confirmed its effectiveness as an instrument of national power—in just 9½ flying days."[22]

Although the sharp knock strategy had to await Richard Nixon's administration, the debate between its proponents and those who favored the progressive squeeze was played out from 1966 to 1968, during the Johnson administration. Joining the president, the Joint Chiefs, and McNamara in the fray was the American public, which was both an actor and a stake in the debate. Public opinion became, in Leslie H. Gelb's phrase, "the essential domino."[23] Johnson was particularly sensitive to the growing public concern. He refrained from the wartime rhetoric which some urged, and yet he knew that dissent over the policy was growing.

McNamara began to have doubts about the effectiveness of the air war in 1966, and by 1967 had himself become a dissenter within the administration. He believed the bombing had reached the limit of its military utility and was fast becoming a political liability. He employed a mix of moral concern and the implied threat of public opinion to argue for a change in policy:

> There may be a limit beyond which many Americans and much of the world will not permit the United States to go. The picture of the world's greatest superpower killing or seriously injuring 1000 noncombatants a week, while trying to pound a tiny backward nation into submission, on an issue whose merits are hotly disputed, is not a pretty one.[24]

The debate on bombing policy began as a conflict over strategy, but it forced to the surface the problem of American ends and means in Vietnam. By the summer of 1967 McNamara was convinced that escalation of the bombing would not be effective. This meant, in turn, that the "unreachable ends" of policy had to be redefined, since further escalation was not politically or morally acceptable. Such a redefinition involved acceptance of the likelihood of a political compromise in the South that would almost inevitably result in a coalition government. McNamara could not convince President Johnson either to restrict the means of the war or to redefine its ends along these lines. However, one month after McNamara left office, in March 1968, Johnson reduced the bombing to the level McNamara had advised, opened negotiations with North Vietnam, and announced he would not run for another term.

The struggle within the U.S. government over how to use American power in Vietnam was a debate about larger themes than the war itself. Policy analysts and plain citizens alike had a sense in 1968 that the "postwar era" of American foreign policy had reached a crucial turn in the road. At stake was whether the policy of containment—of which the Munich and Korea analogies were a part—any longer provided relevant policy guidance in situations like Vietnam. Under question also was whether the United States could, politically and psychologically, sustain a domestic consensus over the kind of limited war which was at the core of the Flexible Response strategy, when the stakes were the political orientation and economic structure of a developing country.[25] The Kennedy Inaugural was optimistic about the capability of the United States at once to relate to the aspirations of the developing world and to renew relationships with European allies. By 1968 the alliance was divided over Vietnam and the frustrations of the war had absorbed much of the energy and resources the United States needed to relate to the rest of the developing world.

The Vietnam trauma did not end in the 1960s. The larger questions it raised about the ends, means, and premises of U.S. foreign policy were still waiting to be answered when the last American was lifted from the roof of the U.S. embassy in Saigon in 1975.

The legacy of the 1960s is a collection of ideas which were shaped then and are used now, policies which have been reshaped or drastically revised, and memories which have yielded "lessons."

It is not surprising that a superpower that regularly stresses the need for stability in the international system manifests a certain degree of continuity in its policies. The influence of the 1960s in the 1980s is more than this expected thematic continuity, however. The continuity is structural in the sense that basic concepts from the 1960s continue to set the terms for policy debate even when the direction of actual policy is defined against them. The staying power of the dominant ideas of the 1960s is notably stronger than that of their counterparts from the 1950s and the 1970s.

President Reagan's intention to proceed with research on the Strategic Defense Initiative, aimed at providing protection against nuclear attack, is an attempt to overturn the relationship of mutual vulnerability which the deterrence policies of the sixties put in place. The Reagan administration's commitment of American forces to Lebanon and its unwillingness to forswear their commitment to Central America has always included the disclaimer that these cases will never become "another Vietnam." The 1960s have left Americans with experiences and concepts which they take for granted in the current policy debate. Yet the 1980s are not the 1960s, and it is important to recognize differences as well as lines of continuity.

The international system of the 1980s still exhibits three tiers: superpowers; middle powers usually in alliance with one of the superpower blocs; and the developing world with a variety of states, some tied to the superpower competition and others seeking distance from it. The growth of international pluralism already evident in the 1960s has become an even more complex phenomenon, however. There are now states which fit the traditional notion of a major power as a consequence of their military capabilities (e.g., the superpowers and China), while other states have become economic powers of the first order but remain militarily weak (e.g., Japan, the OPEC nations, and West Germany). Other nations have sufficient military and economic capacity to participate in key international debates but are not fully independent actors (e.g., Brazil, India, and most of Western Europe); and many others have neither significant military nor economic power, but influence international relations because they are situated in zones of conflict or have become stakes in the superpower competition (e.g., Korea and Angola). Some nations catch the attention of the world not because they have any power but because they are moral catastrophes of world politics (e.g., Cambodia and Ethiopia).

In the midst of this complex web of relationships, the beginning of the 1980s still looks remarkably like the early 1960s, because of the reassertion of the superpower competition. The major powers had not lost the capability to capture center stage in the daily struggle of global politics, but their competition had taken on a more routine character during the latter half of the 1960s and most of the 1970s. The Soviets' enhancement of their nuclear and naval capabilities in the 1970s, when the United States was extricating itself from Southeast Asia, and the subsequent eagerness of the Reagan administration to challenge Soviet power in almost every arena of competition have at times given the 1980s the appearance of a new cold war.

The fabric of the superpower competition is also more complex in the 1980s because of the new role of China. The Sino-Soviet split of the 1960s has moved beyond intramural ideological struggle since the opening of U.S.-Chinese relations in the 1970s and their intensification in the 1980s.

Alliance relationships are more complicated in East and West, though for different reasons. On the fortieth anniversary of Yalta in 1985, the civil struggle in Poland did not threaten Soviet control of Eastern Europe, but it did illustrate the character of the Eastern bloc and the brittle ties which bind it. The Western alliance has posed different challenges of leadership and management. The full-blown emergence of economic interdependence and competition in the 1970s required new efforts at political and economic coordination in a relationship which was originally designed for military and defensive purposes. Even in the sphere of defense policy, the internal political fabric of Western Europe posed obstacles to the Reagan administration's policy of Western rearmament. Echoes of the debates of the 1960s can be heard in all these issues, but the nature of political, economic, and military relationships in the 1980s turns decisions on conventional force expenditures and the deployment of cruise and Pershing missiles into struggles between equals.

The Third World of the 1980s remains relatively impervious to superpower designs but unable to escape the need for outside assistance and the threat of intervention. In Afghanistan, it was the Soviets who found themselves bogged down in a brutal military involvement. The United States was hardly passive about its involvements, supporting covert action in Nicaragua and directing the war effort in El Salvador. Memories of the 1960s placed a fragile but real barrier against the use of U.S. troops in these efforts. The American experience in Lebanon reinforced the distaste for direct involvement in internal struggles, but failed to provide any positive guidance about how to define exceptions to the rule.

Permeating the political and strategic debates of the 1960s was a set of moral arguments that have also left their mark on the foreign policy process of the 1980s. Foremost among these has been the moral assessment of military force in the nuclear age. The classical political dictum that war was the extension of politics by other means had a moral corollary. This held that force could be used, in well-defined circumstances, to achieve morally justifiable ends with morally acceptable means. Nuclear weapons cast the proposition in doubt. Nuclear means seemed destined to corrupt or consume the noblest of ends.

The evolution of strategic doctrine in the 1960s thus produced two questions for the moralist: What was the morality of using nuclear weapons? And how should one assess the morality of deterrence, the threat to use weapons in a strategy of "assured destruction"? Two schools of thought on these questions developed. One position affirmed that the classical moral teaching of a "limited" or "just" war did not apply either to the use of nuclear weapons or to nuclear deterrence. The new weapons demanded a new and distinct moral posture: nuclear pacifism.[26] Most of those who espoused this view still held that some uses of conventional arms were morally acceptable, but they drew the line at both the use and the threatened use of nuclear weapons. A second position in the moral debate emerged, similar to that of the secular theorists of "limited nuclear war" and "counterforce doctrine." This group contended that the moral task was to define the limits of a usable deterrent, one that fit the moral principles on the just use of force, and then to situate strategy within these limits.[27]

Like many of the political categories of the 1960s, so have these moral terms of the nuclear debate been sustained in the 1980s. Two decades of continuing modernization of nuclear armaments have nonetheless shifted the emphasis of the moral argument. In the face of renewed claims by strategists in the 1980s that nuclear weapons could be used with precision and control, there was sharply declining interest in fashioning a moral doctrine for a limited nuclear war policy. Yet the dilemma of deterrence remained. The response of the American Catholic Bishops in 1983 exhibited both continuity with and change from the moral debate of the 1960s. The bishops gave "strictly conditioned acceptance" to the strategy of deterrence while expressing the most severe skepticism about the possibility of any morally acceptable use of nuclear weapons.[28]

This widely discussed assessment formed part of a broader critique of deterrence in the mid-1980s. The organizing idea of the nuclear age—the deterrence strategy—had to come under scrutiny from both sides of the political spectrum. It is, of course, much easier to criticize deterrence than to replace it, but the fact that both the utility and the fragility of deterrence were under public review meant that the defense debate had been driven back to first principles.

A second issue around which moral and political arguments are both highly visible in recent policy debate is that of human rights. While the language of human rights was prominent in the Kennedy Inaugural, it was the congressional involvement and Carter administration policies of the 1970s that placed the issue squarely on the policy agenda. Unlike nuclear issues, which came to new prominence in the early 1980s, the effect of Reagan administration policies has been to push human rights questions to the margin of public debate. The previous visibility of the issue, however, produced an impressive body of analysis of the moral and empirical dimensions of human rights policy.[29] It is a question that cuts across both the superpower competition and the provision of human needs in the Third World. The experience of the Carter years highlighted the difficulty of sustaining a foreign policy that focuses on human rights in a democratic country like the United States. The policy problem has several dimensions: how rights are defined; how the concern for human rights can be incorporated in the definition of U.S. interests; and how human rights can be related to other elements in the foreign policy equation.[30]

The third moral concern that extends from the 1960s to the 1980s is the ethics of intervention. From Vietnam to Central America, the question of what kind of intervention is both justifiable and effective has remained a central element of the policy debate. One result of the Vietnam experience was the conclusion that the United States "should avoid foreign wars not by nipping them in the bud, but simply by staying out of them."[31] By itself this dictum is not a policy, and it surely cannot stand the range of challenges put to a superpower in a volatile international system.

The Vietnam debate was framed principally in terms of the

military means used in the prosecution of the war. It was the bombing and counterinsurgency tactics which sparked the public outcry and which eventually turned key policymakers into dissenters. The argument that it was "costing too much" proved to be the most effective case against the war (though different voices defined these costs in different ways). Since Vietnam there has been some awareness that an adequate ethic and effective policy of intervention must confront questions of ends and means. Thus, the Central America debate of the 1980s, while still inconclusive, has revealed a greater public willingness to question the purposes of the policy as well as the means employed. To press both ends and means issues will force the debate on the ethics of intervention to confront both the complexity of the contemporary international system and the character of the military force which the United States would most likely use in an intervention. In sharp contrast to the extensive moral literature on the nuclear question and the growing resources on human rights, the political and moral analysis of intervention is still inadequate.

To step back from these specific questions and compare the 1960s and the 1980s in more general terms is to encounter one dimension of the meaning of estrangement. It is not an idea that fits easily with the tone or the themes of the Kennedy Inaugural. It is less difficult to relate the concept to the 1980s.

The foreign policy posture of the 1960s was characterized by a clear sense of purpose but not by a solidly grounded view of the world it sought to shape. The world of the 1980s is no less complex than that of the 1960s, yet in a more complex setting the U.S. policy design seems ever more simplistic. The policy pronouncements of the 1980s have not been distinguished by discrimination and nuance.

The dominant category for explaining the world has been the East-West competition. The events of neither decade can be understood apart from the superpower rivalry, but the centrality of this theme hardly makes it a comprehensive category of analysis. The prevalence of this East-West perspective can be attributed principally to the president and his like-minded advisers, but it also may be linked to the broader theme of America's discomfort in today's world.

The thrust of the Kennedy Inaugural and of subsequent policies was to highlight a special role for the United States which required new initiatives to reshape relationships with adversaries, allies, and emerging nations. The initiatives were activist, the tenor was confident, and the goals were ambitious: to refashion previous policies and to undertake entirely new ones in arms control, economic development, and the extension of democratic values. To be sure, the Kennedy policy stressed mutual international burdens and benefits, but it did not consider the possible pitfalls of activism. The unexpected outcome of the 1960s and the events of the 1970s (from new missiles in Russia to new trade deficits at home to new governments in Iran and Nicaragua) have produced a policy perspective that is less inclined to look for new possibilities of action and more concerned about resisting threats to American interests. The reliance on an East-West prism to interpret the world reflects a more pervasive tendency to define problems in terms of "America against the world." International relations have by nature a certain adversarial quality, but the elevation of this theme to the guiding principle of the U.S. policy vision is a recipe for estrangement.

The political and moral consequences of this picture of the world need to be understood in light of an important characteristic of the international system. Observers of international affairs agree that the world has become increasingly interdependent even if they disagree about the policy implications of that fact. Interdependence is a multidimensional phenomenon: The superpowers share a mutual interest in avoiding nuclear war; the alliance relationship poses immediate questions of economic interdependence, from trade policy to monetary issues; and the oil crisis of the 1970s illustrated the advanced economies' dependence on access to Third World resources.

Each of these instances of interdependence is a factual reality today. The essence of effective policy is to provide a vision that gives direction to the facts of life. Increasing factual interdependence is not a policy; it must be shaped by policy. Without policy vision, interdependence can just as easily breed chaos as community in international affairs. A policy vision dominated by adversarial values in an age of increasing interdependence runs counter to both the imperatives and the opportunities of the moment. It fails to seize the positive potential of interdependence, and it increases the likelihood of conflict over growing material interdependence.

We did not have a choice about the fundamental nature of the international system in the 1960s, and we cannot choose our system today. But we do choose which policies project our hopes and designs for the system's reform. The American policy of the early 1960s may have been too ambitious; it had a strong sense of purpose but an insufficient sense of proportion. To set America against the dynamic of interdependence now is to do too little, to set the policy sights too low. Such a posture presumes the privileges of a superpower but refuses to accept responsibility for a more inclusive view of the world. The challenge today is to move from substantive interdependence to political and moral interdependence, from the fact of interdependence to the formulation of rules and institutions for the management of interdependence.

6 Disorder Within, Disorder Without

GODFREY HODGSON

Godfrey Hodgson, a British journalist and historian, has written extensively on American politics, including the highly regarded study *America in Our Time*. Mr. Hodgson has served as Washington correspondent for the *Observer*, the *Times* of London, and the *New Statesman*, has worked in British television, and has been a frequent contributor to the *New York Times*, the *Washington Post*, and the *Chicago Tribune*.

ALL HISTORICAL PERIODS are transitional, but some are more transitional than others. The dozen years from the Cuban missile crisis in 1962 to the fall of Richard Nixon in 1974 was a time of bewilderingly rapid change by any standards. It utterly transformed both the way Americans looked out at the rest of the world and the way the rest of the world, in its infinite diversity, peered warily back.

Although a foreign policy issue—American involvement in Southeast Asia—dominated U.S. politics during a large part of the period, it was nevertheless a time when Americans generally lost interest in the rest of the world. This has been amply demonstrated by polling data. Potomac Associates, for example, asked representative samples of Americans what they thought were the most important issues facing their country in 1964 and again in 1972. Whereas in 1964 half a dozen foreign policy issues, all related in some manner to fear of the Soviet Union, took precedence over any single domestic issue, by 1972 it was the other way around. In 1973, when Henry Kissinger tried to persuade Washington and the rest of the country that there was a danger of confrontation with the Soviet Union over the Middle East, he had great difficulty convincing the media that he was not merely using a foreign policy issue to distract attention from Nixon's troubles at home. Indeed, by the 1970s only one "foreign" problem competed effectively for the attention of Americans with the voices calling for concern with issues of social justice, environmental quality, or domestic tranquility. That one foreign problem was the Vietnam War, which had become such a potent symbol of national perplexity and division that when respondents told a pollster they were concerned about "Vietnam," it was at least as likely that they were thinking about the domestic political schism caused by the war as about "a faraway country of which we know little."

Two presidential inaugural speeches, Kennedy's in 1961 and Nixon's second, in 1973, neatly defined the change in Americans' world view. In 1961 Kennedy did not feel compelled to mention domestic issues at all—not taxes, not health, not education, not civil rights or cities, nor welfare or any of the questions that confounded and obsessed his successor, Lyndon Johnson, only five years later. Nixon, it is true, called on Americans in 1973 to accept the "high responsibility" of building a "structure of peace" in the world. As his term in the White House went on, he did come to rely more and more on his and Henry Kissinger's showy feats in foreign affairs to distract the country from less palatable events at home. Yet, in that second inaugural speech Nixon made it unmistakably plain that "meeting our challenges at home" was the first priority. The contrast was not only between the values and temperaments of two presidents, different as these were, but also between the national priorities of 1961 and 1973.

In retrospect, it also looks as if the sense of danger that caused Kennedy to harp on the "missile gap" during and after the 1960 campaign, and made many Americans feel threatened by successive crises in Berlin, Cuba, the Congo, and Laos, was exaggerated. This sense of vulnerability was the other side of the coin of excessive national self-esteem. Because Americans unrealistically expected the world to conform to their expectations, they were outraged when it did not. In the early 1960s, as Daniel P. Moynihan pointed out later, Americans gave the impression that they were concerned with the outside world largely because they believed their own society was not in need of attention.[1] By the mid-1970s they had no alternative but to turn their attention to insistent problems at home. Complacency was shattered. In that brief span the United States traveled from the Augustan assurance of mid-Victorian England to something closer to the angry confusion and *Weltschmerz* of Weimar Germany. While the fires of black protest were burning ten blocks from the White House, and the intelligentsia were prophesying the dissolution of American society in every weekly and monthly magazine, somehow the problems of the Congo and even of Cuba seemed less urgent than before. Between the assassination of Kennedy and the disgrace of Nixon, Americans became less willing to shoulder the world's burdens, more cautious about serving as its banker, policeman, or even night watchman.

This swift change in the national mood evoked a parallel transformation in the outside world's perception of the United States. If nothing succeeds like success, defeat, as John F. Kennedy liked to say, is always an orphan.

On the morrow of the October 1962 Cuban missile confrontation, during the winter of 1962–63, America's most obstinate critics

—even Charles de Gaulle—were impressed by the way the United States had passed the test, imperial in its strength and restrained assurance. A dozen years later even those foreign observers who knew the United States best and admired her most were shaken. The favorite epithets seemed to be "unsteady," "unpredictable," "perplexed," and "neurotic." To America's friends, this sudden loss of confidence and purpose was disconcerting. To rivals, it was an opportunity. To adversaries, it was an irresistible temptation. The turmoil in the Western alliance, and the new adventurism in Soviet policy, clearly visible from 1974 on, can be attributed in part to the domestic political upheavals in the United States.

The drama of that momentous decade was not the result merely of some isolated events called Vietnam or Watergate. Rather, the interaction of three vast, extremely complex historical processes, each a long time in gestation, produced a sudden acceleration of change.

The first of these processes was the domestic rebellion and self-questioning touched off by the civil rights movement. Its impact dramatically altered the expectations and, to a lesser extent, improved the condition of black people in the United States. But it also went far beyond them.

The second was the global commitment to resisting communism, by force if necessary, of which the war in Southeast Asia was a test case.

The third was a crisis of government, whose origins date back to the beginning of American history, but which became acute when Lyndon Johnson succeeded John Kennedy, setting off a chain of events that culminated in the downfall of Richard Nixon. The U.S. Constitution had so carefully checked, balanced, and separated executive and legislative power that it constrained the development of an effective modern presidency. In the extraordinary circumstances of the Great Depression and World War II, the equally extraordinary abilities of Franklin D. Roosevelt made such a presidency possible. From Harry Truman's time on, however, presidents have found themselves stalemated; Johnson, Nixon, Ford, and Carter were all driven from office in frustration.

The combined effect of these three factors—domestic turmoil, an unsuccessful foreign war, and governmental paralysis—would not only render American leaders, for the better part of a decade, nearly impotent to carry out their designs for the country's well-being, but would also challenge virtually all the key assumptions

Americans had made during the brief period of unquestioned success and supremacy after World War II. In particular, all the prevailing assumptions about the U.S. role in the world were to be harshly called into question.

In October 1962, at the height of the Cuban missile crisis, Kennedy dispatched the senior living American diplomat, Dean Acheson, on a confidential mission to brief first French President Charles de Gaulle, and then the leaders of NATO, on his plans for a naval blockade of Cuba. Acheson flew to France overnight in a military aircraft, and on the morning of October 22 hurried to the Elysée Palace to see the general. He showed him aerial photos proving the deployment of Soviet missiles and bombers in Cuba and handed him a letter from Kennedy. De Gaulle reassured Acheson. "You may tell your president," he is reported to have said, "that France will support him." Afterwards the Gaullists made much of this declaration of solidarity. The differences between the United States and France might be many and serious, they implied, but when the hour of danger sounded, de Gaulle and France knew where their loyalties lay. When the Soviet ambassador to France, Sergei A. Vinogradov, crudely threatened the general with the consequences of incurring Soviet wrath, de Gaulle dismissed him with magnificent aplomb: *"Eh bien, Monsieur l'ambassadeur, nous mourrons ensemble."*[2] ("Oh well, Mister Ambassador, then we will die together.")

But de Gaulle made another point to Acheson at that meeting. According to some reports, he deplored the fact that he had been informed, rather than consulted, by Kennedy. (In other versions, de Gaulle is said to have asked Acheson whether he had come to inform or to consult.) De Gaulle saw the relationship between France and the United States without sentiment or self-deception. He also believed that the Soviet leader, Nikita Khrushchev, had been bluffing when he deployed missiles to Cuba. Before long de Gaulle grasped that although the Cuban missile crisis had ended in a temporary triumph for the United States, the humiliation of the Soviet Union offered France an opportunity to pursue her own policies and interests independently of the Americans.

De Gaulle never revealed his precise analysis of the Cuban missile crisis. But he recognized that Khrushchev and the rest of the Soviet leadership, having been forced to retreat with nothing to show for the Cuban adventure, would not soon be in a position to put serious pressure on the United States or on Europe: Khrushchev himself never recovered politically from the Cuban gamble. Thus, it would be safe, if only temporarily, for Europe to act more independently, without serious danger of a Soviet offensive. Paradoxically, however, the crisis had also revealed the weakness, almost the irrelevance, of a Europe whose fate could be determined over its head by the two superpowers. De Gaulle realized this, and so, in the aftermath of the Cuban missile crisis, he saw both an opportunity and a need for a policy of *grandeur*.

Over the next four years de Gaulle struck a number of shrewd blows against what he considered the Anglo-Saxon domination of the Western alliance, and therefore of the world. In the same famous press conference in January 1963 during which he haughtily dismissed Britain's request for membership in the European Community, he also made explicit his perception of Europe's danger—and Europe's opportunity. "In the light of the Cuban affair," he said, "American nuclear power does not necessarily and immediately meet all the eventualities that concern Europe and France."

In 1964 de Gaulle's blows fell with an accelerating rhythm: his recognition of the People's Republic of China, followed by visits to Mexico and South America that were presented as challenges to an American sphere of influence. In 1965 came the verbal assault on the Bretton Woods system of postwar international finance and trade, which had been sponsored by the United States and Britain. Soon the French government launched a speculative attack on U.S. gold reserves. At the same time, de Gaulle ordered a boycott by France of most of the institutions of the European Community (a move he had to rescind quickly, when it became apparent that it hurt France much more than her partners).

While de Gaulle was systematically sapping the position of the United States in the world wherever he could, he was also beginning to court the Communist powers. In February 1965, at the very moment when Johnson was planning the first bombing attacks on North Vietnam and the dispatch of the first U.S. ground troops to South Vietnam, de Gaulle sent a letter of sympathy to France's old antagonist in Hanoi, Ho Chi Minh. His recognition of China was admittedly a perfectly rational step that not only gave France certain commercial and diplomatic advantages but may also have helped prepare the way for the Nixon-Kissinger opening to China eight years later. But by May 1965 France was voting alongside the Soviet Union at the United Nations to condemn LBJ's invasion of the Dominican Republic. And the next year de Gaulle's flirtation reached a climax with his grand visit to Moscow.

In the spring of 1966 de Gaulle ordered France's withdrawal from NATO, though not—as he was careful to make plain—from the North Atlantic Alliance. That summer, in the course of a trip around the world, he chose a visit to the doomed city of Phnom Penh, Cambodia, to issue his most immoderate diatribe against American policy, calling it "ever more menacing to world peace." Meanwhile, General Charles Ailleret put the theoretical capstone on de Gaulle's structure of proud and prickly independence by announcing that the French nuclear strike force stood ready to defend France *à tous azimuts,* at all points of the compass.

By 1968 the pretensions and the inner weaknesses of de Gaulle's scheme for restoring the splendor and the independent strength of France had been cruelly exposed. De Gaulle's grand design exaggerated France's power, and his conception of a reunited Europe stretching from the Atlantic to the Urals was a fantasy. His international ambitions seemed all the more unrealistic after the French worker–student revolt of the spring of 1968 shattered his standing at home. Yet de Gaulle's five-year spree was not mere folly and self-indulgence. Nor was he simply taking advantage of the American preoccupation with Vietnam. De Gaulle's diplomatic campaigns achieved a measure of success because they were grounded in the new realities of the decade after the Kennedy assassination.

The relative power of the United States, in economic and strategic terms, had already sharply declined from the peak it had reached after World War II. The rest of the world had never really accepted America's postwar hegemony as normal or legitimate. De Gaulle was hardly the only world statesman hankering to tilt the balance of power back to a state where the United States no longer enjoyed political primacy. Most Americans thought their country's supremacy during the immediate postwar years was now the natural state of affairs; most Europeans tended to believe that this was no longer deserved.

It was beyond Charles de Gaulle's personal power either to return matters to what he regarded as a more normal condition or to persuade the American public that it was wrong. There were elements in the general's bitter struggle to reduce Anglo-Saxon predominance that were purely personal, derived from his experience with Roosevelt and Churchill in World War II and from earlier traumas. But without the edge of personal pique, much that de Gaulle was trying to achieve later came to motivate far less anti-American leaders like West German chancellors Willy Brandt and Helmut Schmidt. The course of American foreign policy during the decade after John F. Kennedy's assassination is nevertheless the story of, first, a painful adjustment to the diminished relative status of the United States and, subsequently, an overreaction by the American public and policymakers to this adjustment.

The troubled period through which American society passed between the middle 1960s and the middle 1970s was caused only in part by foreign issues. It is true that the Vietnam issue became the focus of anger and disillusion, yet it is also true that the Johnson administration, by grossly underestimating the domestic political and economic costs of the war, contributed to the wave of dissent and discontent. For a time this discontent focused mainly on the unwinnability, the unpopularity, and the perceived immorality of the war, but it really went much farther.

If this domestic crisis was not wholly or even mainly caused by international events, its consequences for American foreign policy were nonetheless profound. In retrospect, it may be that the pendulum swung too far. While questioning the excesses of the imperial America of the postwar period, some critics also questioned much that was sound, or natural, or at least inevitable, in U.S. domestic and foreign policy. As a result, the decade from 1963 to 1973 can be divided with a neatness that is rare in history. In the first half of the period, from 1963 to 1968, the crisis of confidence was so shattering that the forces of challenge and change went almost unchecked. The fire on which so many of the assumptions of the American consensus had been stacked blazed out of control, and for a time it looked as if its sparks might set afire the whole structure of traditional American values and institutions.

That did not happen, however. For one thing, the nature of the movement against the war had been misunderstood. By 1968 the polls indicated that a majority of Americans opposed the war, but that opposition was made up of groups with different and, in some respects, contradictory views.[3] The "peace movement"—those who thought the war was criminal—were joined in the opinion polls, and sometimes in the voting booths, by those who merely thought it a mistake. The civil rights movement, having achieved its initial goal of formal desegregation but having discovered how much more difficult it was to achieve actual equality, began to lose its political impetus. Disillusion set in with the program of liberal reform offered by Johnson and other regular Democrats. And a decade of unbroken prosperity began to ebb with the onset of inflation in the second half of 1965.

The turning point came in 1968.

Up until that year it looked as if the initiative lay with people intent on pulling apart the central assumptions that had guided American society and American policy toward the rest of the world since the New Deal. But Lyndon Johnson fell and was replaced not by the wild-eyed prophets of a new heaven on earth, but by the embodiment of nationalism and Middle America, Richard Nixon.

In some respects the changes in American society were even more far-reaching from 1968 to 1973. The "women's movement," for example, challenged prevailing assumptions more deeply than any other single political phenomenon of the time. Still, while those who sought to sweep the liberal consensus away and replace it with radical alternatives did not come out of the jungle with their hands held over their heads in surrender, nonetheless, after 1968, the initiative steadily passed to those who felt that the pendulum of change had gone beyond reasonable limits. Reaction began to replace rebellion.

What was at stake in all this was nothing less than the continuance of the America that had emerged in 1945. Much becomes plain once it is grasped that while to a majority in the United States the brief, unchallenged American superiority of 1945–55 seemed normal, to much of the rest of the world it had always appeared to be an aberration. Now it looked, and not only to General de Gaulle, as if that aberration was about to be corrected.

It was only in the last year of World War I that the United States emerged for the first time as a power of the first rank. For a brief period, with Europe prostrated by the war, America seemed in a position to dominate, but the American people were not then ready for a leading role. The Senate disowned Wilson's internationalist strategy. Missionary ambition was replaced by isolation from European affairs. American military power subsided, and the industrial expansion of the 1920s was blunted by the Great Depression. Before America could be in a position to triumph over Hitler's Germany and imperial Japan, it had to convalesce from a debilitating decade of economic crisis and social hardship. The war itself hastened the convalescence.

After World War II the situation was quite different. The United States was ready for greatness, and its only serious rival was the Soviet Union, whose aggressive stance under Stalin was in part a bluff meant to conceal its exhaustion.

American military power at that time seemed almost casual, an accidental by-product of an economic machine whose real function was to churn out the goods and services that would allow more and more U.S. citizens to attain the good life they had aspired to. The relative dominance of the American economy in those brief years of the would-be American Century had never been approached before, certainly not by Victorian Britain in the almost equally brief period when it enjoyed a head start over its rivals in the early stages of the Industrial Revolution. For a few years, beginning in 1950, the United States possessed a near-monopoly in the production of all the primary components of military power: steel, oil, rubber, aircraft, automobiles, and electrical manufactures, not to mention scientific and research potential. As vast as the American preponderance of all the sinews of conventional war might be, however, it was less decisive than the most precious monopoly of all, in atomic power.

This unprecedented power was accompanied by a will to use it. After 1945, and more particularly after the crucial period of Cold War commitment under Truman and Acheson in 1947–50, the United States was prepared to intervene in Europe, in the Middle East, in Latin America, and in the Orient. It did so to resist communist encroachment, to help its friends, or to advance its own political or economic interests. Sometimes in alliance with European colonial or corporate enterprises—but sometimes, too, in rivalry with them, as in the deliberate exclusion of Britain and France from the Saudi Arabian oil fields—the United States was laying the foundation for a worldwide economic empire that would pay back the cost of the Marshall Plan and all the rest of the foreign aid of those years put together. American corporations were buying their way into Middle Eastern oil, South American iron ore and copper, African metals, Indonesian and Malaysian tin and rubber, Canadian and Australian agriculture, and European banking, manufacturing, and service industries. The scale of this endeavor made the largest previous export of capital, by the British, the

French, and the Germans in the period immediately before World War I, look insignificant by comparison.

Power politics and power diplomacy might go against the American grain, yet there was at the time little reluctance to assume the mantle of world leadership, and little opposition. Most Americans thought there was no tolerable alternative, when Stalin's Russia seemed to be actively threatening every society in the world. Many of the prewar isolationists were fiercely anti-communist, so they were torn; but what quite a few of them wanted now was not isolation but bolder and more effective intervention, only by preference in Asia rather than Europe. The crucial time of decision came in the summer of 1950, with the promulgation of NSC 68.[4] Congress acceded, almost without discussion, to a doubling of the defense budget and of military personnel. The Truman administration threw itself not only into defending South Korea, but also into the creation of a worldwide military and alliance system, with bases from Saudi Arabia and the Philippines to Greenland, greatly expanded intelligence and covert-action capabilities, plus increased nuclear weapons production, naval procurement, and strategic bomber manufacture.

This commitment to global power,[5] with all it implied for doctrine, for investment, and even for a certain degree of militarization of American society, was not the policy of one political party or any particular section of public opinion. It was all but unanimously accepted among what were acknowledged to be the most enlightened and influential elements in the nation. For a start, globalism was the policy of the presidency from Franklin D. Roosevelt's time through the Truman and Eisenhower administrations into those of Kennedy and Johnson. It was strongly supported by the newly emerging national media, including, when they came on the scene in the 1950s, the television networks. Preeminently, it was the policy of the great pool of talent clustered around the big New York banks, the Wall Street law firms, and the major American graduate schools which, with the professional military, provided a high proportion of those who formed and managed U.S. foreign policy—the so-called foreign policy establishment.

In these overlapping circles of influence, from 1950 until 1965, a commitment to the assertion of American global power was the badge of membership. To question globalism—or, as its adherents skillfully named it, internationalism—was to risk exposing oneself as a know-nothing, a provincial, or a reactionary. And so a foreign policy consensus came into existence, mirroring the larger consensus that had held the middle ground in American politics since the United States entered World War II.

American foreign policy was thus set in concrete during a brief, atypical period, when the United States enjoyed not just a nuclear monopoly but something close to a monopoly of all the other ingredients of power. A generation of Americans had grown up expecting the world to conform to their ideas. A generation of American leaders had emerged who believed it would be possible to intervene effectively around the world at comparatively acceptable levels of cost as the United States had done in Lebanon and Guatemala in the 1950s. And a generation of leaders had grown up elsewhere, including Western Europe, who were inclined to accept the inevitability, even the desirability, of American dominance.

Once a synergy began to operate between war in Southeast Asia, domestic dissent and doubt in the United States, and the malfunction of the American political system, it was natural that most of the assumptions of the previous consensus should be questioned and widely rejected. That consensus embraced the main lines of both dometic and foreign policy. One of the first victims was the assurance that had characterized foreign policy. By the early 1970s Henry Kissinger might be dazzling the world with his cloudy theoretical constructs and his tricky diplomatic footwork, but the old unity of purpose had disappeared, and so, too, had much of the world's confidence that Washington knew what it wanted and knew how to get it.

Consensus, of course, is relative. To the men who gathered in Washington at Christmas in 1960 to serve Kennedy, the break between their politics and those of the outgoing Eisenhower administration seemed sharp. From a longer perspective, however, the continuity is more striking than the distinctions. What happened in 1961–63 now looks like a reformation, rather than a transformation, of a foreign policy whose essential features had not changed. It was later, after Kennedy's death and Johnson's entrapment in Vietnam, after the turmoil of the late 1960s, that the break came.

The Kennedy men saw themselves as innovators. They meant to end what they perceived as a period of disgraceful lethargy and vacillation, a period when—in the phrase borrowed from St. Paul by their favorite military man, General Maxwell Taylor—"the trumpet gave an uncertain sound." They came armed with new, more subtle strategic doctrines; the watchwords, made fashionable by the defense intellectuals in Cambridge, Massachusetts, during the late 1950s, were "flexible and graduated response." They saw themselves—which newcomers to Washington do not?—as more vigorous and effective than their predecessors and their allies. "I don't expect," said President Kennedy on New Year's Day 1963, "that the United Sttaes will be more beloved, but I would hope we could get more done."

In retrospect, these were mere changes of nuance and tone from the Eisenhower years. By far the most important difference in substance, which was usually missed by those who were beguiled by the administration's brilliant presentation of its own image, was what Kennedy's chief speechwriter, Theodore Sorensen, later called the "build-up of the most powerful military force in human history —the largest and swiftest build-up in this country's peacetime history." In accordance with the theory of graduated response, the president armed himself with everything from "the most massive deterrents to the most subtle influences." In plain language, that meant more Minuteman and Polaris missiles, more ships and infantry divisions, and a greatly increased capability for intelligence gathering, covert action, and counterinsurgency. The Kennedy administration's obsession with this hairy-chested form of warfare contributed to a fatal overestimate of what the United States could hope to achieve in Southeast Asia.

Neither the arms buildup nor the professed willingness to fight the forces of revolution in every jungle and barrio of the Third World constituted a sharp break with the policies of the Eisenhower administration, which had, after all, commissioned both Minuteman and Polaris, overthrown radical governments from Iran to Guatemala, and fought a miniature Vietnam War on behalf of Ramon Magsaysay in the Philippines. The shipwreck at the Bay of Pigs put an end to a bipartisan enterprise.

The foreign policy consensus had its origins in World War II, when the American people and their leaders decisively rejected isolationism, leaving only insignificant, if resentful, minorities in disagreement with the proposition that the United States must intervene in the world to protect its interests—one of which was the survival of democracy in Europe—and to prevent the emergence of any unacceptable world order. The second stage came with the Marshall Plan and the Truman Doctrine, reaffirming that commitment in the face of Stalin and his challenge in Europe. And the third stage, already implied on a theoretical level by the Truman Doctrine, was the global commitment that came as a response to the crisis in Korea.

Throughout the 1950s and into the Kennedy years the main lines of this foreign policy were accepted by a consensus, in the strict sense of that word; that is, they were rejected only by a handful of pacifists on the Left and a scarcely more significant rump on the Right who called for the roll-back of communism.

It was in 1963 that the national foreign policy consensus began to break down. This was most obvious, of course, in the case of the U.S. involvement in Vietnam. The fall of President Ngo Dinh Diem, only a few weeks before John F. Kennedy's assassination, marked the beginning of the end of the illusion that South Vietnam could survive without massive American help. Viewed with hindsight, there is an ominous inevitability about events leading to Lyndon Johnson's decisions to commit U.S. air power and ground troops early in 1965. In his State of the Union message the following January, Johnson for the first time explicitly placed Vietnam where it had already been for two years, and would remain for almost a decade—"at the center of our concerns." But he insisted that "this nation is mighty enough, its society is healthy enough, its people are strong enough, to pursue our goals in the rest of the world while still building a Great Society here at home." One day Johnson may be proved right that America can accomplish everything at once, but in his own time he was wrong. Increasingly over the next two years he had to turn his attention away from his domestic program, as the war consumed his titanic energy. That was not all it consumed. It soon became clear that Johnson had gravely underestimated Vietnam's economic costs. Mighty the United States unquestionably was, but not mighty enough to win that particular war without paying a price, in several currencies.

The effects on foreign policy of the civil rights upheaval and the crisis of government were less direct, but still significant. Until 1963 the civil rights movement was widely seen as having an essentially benign effect on American society. The March on Washington that summer seemed to be the penultimate stage in the elimination of a residual historical problem, legally enforced segregation in what looked from New York and Washington like a backward corner of the nation. But that same summer it became clear that, as fast as the demand by southern blacks for legal equality could be met by legislation, blacks in the North would demand their own economic and social transformation. If this were to come at all, it could only be gradual and, moreover, at the expense of other groups in society.

The international consequences of this dramatic and painful upheaval for the United States were important, but contradictory.

Twenty years after his death, John F. Kennedy's photograph, torn from some aging news magazine, still hangs in many sweltering hovels around the world. It is no disrespect to Kennedy's achievements to say that his near-canonization in many parts of the Third World was based more on emotion than on an accurate knowledge of what he stood for. People in Africa, Asia, and Latin America had heard—and possibly felt the benefits—of the Peace Corps, food aid, and the Alliance for Progress. They had also heard that people with non-white skins like their own had been badly treated in the United States and that President Kennedy was against that. He appeared to them, therefore, as a great and good man, the friend of the poor and the enemy of their enemy, colonialism. It is unfair that in those same parts of the world, Lyndon Johnson, who was committed to taking action on behalf of the civil rights of blacks both earlier and with more understanding than Kennedy, should be remembered primarily as the heartless bully who ravaged Vietnam, while Kennedy, who was at least as responsible for American involvement in Vietnam, should be remembered as an angel of light.

Kennedy's death and his replacement by Johnson ended a brief period when the United States was more popular in the Third World, with the elites and the masses alike, than at any time since the immediate aftermath of World War II. As the United States became more committed to using its wealth and its firepower to defeat what was generally seen in the former colonial countries as a movement for national independence in Vietnam, anti-American feeling spread and intensified. Local politicians of many persuasions fanned the flames for their own purposes.

In the developed countries the social upheavals in the United States weakened American prestige and influence in a quite different way. Black discontent proved to be the model for a whole series of other protest movements. The antiwar movement, the student movement, the environmental movement, native Americans, Chicanos, gays, even some elements of the women's movement—all of these borrowed from the style and the rhetoric that had been forged by black activists. The rhetoric of dissent was much given to exaggeration, designed to raise the consciousness and bolster the morale of each group. But this same exaggeration was often taken all too literally abroad. Readers in Munich or Milan, still less Moscow, could not be expected to discount the tactical purposes of hyperbole. Learning that a prominent black leader believed that the United States was in the grip of fascism, or that the country was on the brink of revolution, or, as one much-admired professor wrote in those strange days, that "America's history as a nation has reached its end," Europeans could be forgiven for wondering whether something had indeed gone very wrong with the America they knew and respected.[6]

As the cacophony reached its bizarre climax in the election campaign of 1968, it was precisely those solid, cautious circles in Europe that had relied on American steadfastness that were most disconcerted, then frightened, then finally annoyed. British Conservatives, German and Italian Christian Democrats, and other conservative elements in Western Europe—conspicuously including the military, most industrialists and bankers, and the readers of such journals as the *Economist*, the *Frankfurter Allgemeine Zeitung*, and the *Neue Zuricher Zeitung*—had looked up to the United States as a society free from the unruly and disruptive elements that were all too familiar in Europe. What were they now to believe, as students trashed the Harvard and Columbia campuses, protesters in California tried to block troop ships bound for Vietnam, and blacks burned stores within ten blocks of the White House? Was this the last best hope of sanity, order, and conservatism?

The crisis in American government affected overseas opinion toward the United States in much the same way, though more slowly. The Kennedy assassination was the first shock to Europeans who had regarded the United States through the prewar rise of fascism, the war, and the postwar period as a haven of civility and sanity. The arrival of Lyndon Johnson in the presidency did not help. It was all too easy to see him, as indeed many American commentators did, as an uncouth, rough-spoken throwback to a cruder style of American politics. This was a caricature of the complex personality of one of the ablest American political leaders of the twentieth century. But, fair or not, it was a stereotype that diminished the prestige of the United States in the late 1960s. The riots, the sit-ins, the subsequent assassinations of Martin Luther King, Jr., and Robert Kennedy, the turbulence of the 1968 election—all reinforced the impression that the United States was in an unsettled state, that its government could not be relied on for steady, consistent behavior.

Then Johnson was succeeded by Nixon. In time the foreign policy experts and professionals in Western Europe and elsewhere—notably including Moscow and Peking—came to hold a relatively high opinion of Nixon, derived only partly from their even higher opinion of his national security adviser, Henry Kissinger. To European media and public opinion, however, the new American president was the old "Tricky Dick," the man from whom you would never think of buying a used car, the persecutor of Alger Hiss (generally seen in Europe as a martyr to right-wing intolerance), the anti-communist bigot, and the blue-jowled enemy of the sainted Kennedy.

But the damage done to the American image by the spectacle of Nixon acceding to the White House was nothing compared to the devastation wrought by the bizarre revelations that began to emerge in early 1973. It did not matter that some of the things the president was accused of—for example, taping conversations in the Oval Office—his predecessors were known to have done as well. Nor did it matter that other abuses—bugging telephones, for instance—were widely practiced in Europe and were not even illegal in many countries. The overall effect of the sequence of events known as Watergate was to diminish the credit of the U.S. government.

What Watergate suggested to some foreigners was that Nixon and his cronies were trying to build a police state. To others it merely demonstrated their incompetence. Virtually everywhere, the role of the American media in all this seemed at once admirable and odd, evidence of the lack of seriousness of American politics. How could anyone trust a government that could not even keep its secrets from the *Washington Post*? The editor of the *Times* of London, Sir William Rees-Mogg, flew to Washington at one point to reproach the media for behaving like a "hunting pack." And when, during Nixon's last desperate presidential tour of Europe and the Middle East in 1973, American reporters began to ask extremely pertinent questions about Kissinger's involvement in the Watergate improprieties, the European press, in almost unanimous chorus, deplored the idea of dragging foreign policy through the mud of domestic scandal. It did not matter that many of the most serious charges raised against the Nixon White House, including the break-in at Daniel Ellsberg's psychiatrist's office, arose precisely out of the administration's concern with foreign policy secrecy, nor that many of these matters seemed to originate suspiciously close to Kissinger and his ambitious assistant, Alexander Haig. The cry in Europe was one of relief: "Thank goodness Henry's not involved!" said the secretary of state's sophisticated admirers.

Few in Europe even attempted to understand the role of Congress or the courts. The tripartite separation of powers in the U.S. Constitution is not common elsewhere, and the role of the judiciary and the legislature in uncovering and punishing misbehavior in the executive branch was widely taken as evidence not that the American system had worked as intended, but that people in Washington had gone out of their heads, and so it was no use trying to transact serious business there for the duration.

By 1974, as a result of the cumulative damage done by all these circumstances, and especially by the culminating blow of a president's forced resignation, American foreign policy was dead in the water. Kissinger has described in his memoirs how already in the spring of 1973 "the symptoms of weakening authority were everywhere." Chinese officials were discreetly checking out their theory that organized groups in the United States were orchestrating the opposition. And, according to Kissinger, Leonid Brezhnev cut short his American trip in 1973, either because he wanted to underline the "erosion of executive authority" in America or perhaps even because he genuinely believed that the United States was on the brink of a dangerous crisis. Kissinger, of course, had his own reasons for exaggerating the damage done to American foreign policy by his opponents, and for emphasizing how vital it was that they not inquire too closely into his own role in the scandals. His memoirs are an exercise in self-exculpation. Yet he is surely right in contending that the crisis of government that drove Nixon from office crippled U.S. relations with allies and adversaries alike for several years.

Kissinger portrays Watergate as an unprovoked assault on a statesmanlike president and his devoted helpers by their political enemies. In reality, Nixon's fall was the result of his own arrogance and folly and that of his associates, including Kissinger. In particular, it was a consequence of the secrecy that was both an obsession and a necessity for the Nixon administration. Nevertheless, it is hard to quarrel with Kissinger's conclusion that

> with every passing day Watergate was circumscribing our freedom of action. We were losing the ability to make credible commitments, for we could no longer guarantee congressional approval. At the same time, we had to be careful to avoid confrontations for fear of being unable to sustain them in the miasma of domestic suspicion.

Unfortunately, the suspicion was abundantly justified.

Kissinger has a remarkable gift for rationalizing the unavoidable. From time to time—beginning with his incessant briefing of favored journalists in Washington, not all of whom were able to understand what they were being told—he has employed grand strategic conceptions, presenting his actual or preferred policies as the logical implementation of this or that theory or doctrine. The reality is perhaps the other way around: He is a fluent escapologist who finds grandiose justifications for each improvised dodge or sidestep. One should not be too condescending about that. International affairs are too unpredictable for even the most brilliant statesmen to see very far into the future. Still, it is important not to take Kissinger's theoretical constructions any more seriously than he does himself.

America's real problem in the period under discussion was the Vietnam War. The origins of the Nixon-Kissinger foreign policy are best seen, perhaps, through the eyes of Richard J. Whalen, the foreign policy adviser and speechwriter who resigned from Nixon's staff in the summer of 1968. Whalen described the evolution of Nixon's thinking about Vietnam in his book *Catch the Falling Flag*.[7] By the spring of 1968, he writes, candidate Nixon was convinced that the war could not be won, but he did not dare admit this openly to the American people. That, Nixon supposed, would be political death for a conservative Republican. What Nixon said to Whalen was this: "I've come to the conclusion that there's no way to win the war. But we can't say that, of course. In fact, we have to say the opposite, just to keep some degree of bargaining leverage."

Nixon's first impulse, Whalen said, was to make a campaign speech calling on the Soviet Union to help negotiate a settlement in Southeast Asia, but he never gave that speech. Hours before he was due to tape it, Lyndon Johnson announced that he had decided not to stand for reelection. The political landscape was transformed, greatly to Nixon's advantage, and caution told him not to take any chances.

The problem, Nixon and Kissinger seem to have understood, was how to reconcile two apparently irreconcilable propositions. On the one hand, the war could not be won. On the other, it could not be seen to have been lost. Nixon saw the issue entirely in terms of domestic politics; a president or a presidential candidate who admitted that the war could not be won might be punished at the polls. But Kissinger probably saw the problem more in terms of preserving the country's diplomatic credibility. Their reasoning may have differed, but in any event Kissinger, especially in his first years in the White House, displayed a strong inclination to agree with anything Nixon seemed to think.

What was to be done? The solution they eventually hammered out was not a simple one. It involved intricate diplomatic maneuvering and a subtle mélange of graduated violence and psychological warfare in Vietnam, combined with calculated deception of enemies, allies, and the American people alike. The strategy frequently had to be modified, and at times the sheet music was laid aside in favor of sheer improvisation. But, ultimately, their policy was not so complicated after all: The United States would have to withdraw from its commitment in Southeast Asia, but gradually, even surreptitiously. (Nixon and Kissinger were not bold enough to adopt openly the advice of Republican Senator George Aiken of Vermont, which was to "declare a victory and bring the boys home." Yet what they were trying to do, on a much longer time scale, was no different.)

There were three elements to this strategy of withdrawal, as the president and his national security advisers worked it out in 1968–69. There would have to be negotiations with the North Vietnamese. The South Vietnamese would have to take over a growing share of their own defense. And, to avoid the accusation that the Nixon administration was abandoning an ally too precipitately, South Vietnam's odds would have to be secretly improved with clandestine bombing of the so-called Ho Chi Minh trail, which the North Vietnamese used to supply their troops in the South, and of the Viet Cong's sanctuaries in neighboring Laos and Cambodia.

The first element—negotiations—involved pressuring the North Vietnamese through their friends and suppliers in Moscow, rather than searching for neutral intermediaries, as the Johnson administration had fruitlessly done. This tack required secret diplomacy with both the North Vietnamese and the Russians. In a novel effort to push Moscow to push Hanoi, a third line of secret diplomacy was opened with Peking. Over time, extraneous issues were brought into the secret discussions, first with Moscow and then with Peking. Among them were such important matters as the strategic arms limitation talks, the sale of grain to the Soviet Union, and the opening of new relations with China. Kissinger proved adept at the intrigues involved, and a new style of American diplomacy emerged. It was ironically similar to the old, sly European technique, long rejected by America in favor of "open covenants openly arrived at." (The new style did not, however, long survive Watergate and the fall of Nixon.)

"Vietnamization," the second element of the Nixon-Kissinger strategy, also involved a degree of deception, since South Vietnamese governments could hardly be expected to hold firm if they realized that the United States intended gradually to remove its support. But, to keep Saigon in line, it became necessary to deceive the American people about how long South Vietnam's resistance, once "Vietnamized," could be expected to survive. In the process, Vietnamization was universalized, becoming the model for a Nixon Doctrine under which the United States, acknowledging the impossibility of acting everywhere at the same time, would instead maintain a *Pax Americana* through regional surrogates.[8]

The theory had a plausible ring to it. The relative strength of the United States had declined significantly since the years when the Truman Doctrine was proclaimed, implying a global commitment. Kissinger himself pointed out that the gross national product (GNP) of the United States, about half that of the whole world in 1950, had declined relatively by about 10 percent each decade, so that by the 1980s America would account for less than 20 percent of world GNP. Adversaries and potential adversaries had grown stronger, but so had friends and allies. It was not just more difficult for the United States to guarantee the security of countries like South Vietnam; there was less reason why the United States should do so alone.

The way the Nixon administration proposed to have friendly countries share the burden, however, was very different from the method adopted by the Eisenhower administration. Instead of a girdle of regional alliances—NATO, the Central Treaty Organization (CENTO), and the Southeast Asia Treaty Organization

(SEATO)—Nixon pursued a series of bilateral relationships with congenial regional powers: South Vietnam, Israel, Turkey, Pakistan, South Africa, Iran. Most of these countries chosen for the honor of helping the United States maintain order in the world had governments of the authoritarian right. Indeed, there were clues from Kissinger that the Nixon Doctrine had a more overt ideological bent than the Truman Doctrine. Truman had pledged to help any government threatened with external aggression. The Nixon Doctrine, ostensibly a "retreat of American power," often looked uncomfortably like a pledge to protect governments from domestic political opposition of the left, as well as from external dangers. Surrogates like the Shah of Iran were to be heavily armed for conventional defense, and then the United States would back him and others with money, diplomatic support, covert action, and the ultimate threat of strategic air power.

The doctrine was applied erratically. Kissinger was ready to intervene to prevent left-wing forces from taking power, or holding on to it, in Chile and Angola. The United States was at least as involved in the world's conflicts as in the Kennedy and Johnson era. This time, ironically, it had fewer means at its disposal, and often, because of the solitary way Kissinger operated, an inferior institutional understanding of various conflicts. Kissinger's indifference to the Bangladesh crisis of 1971, for example, led to a bloody war and a lasting loss of influence for the United States in the Indian subcontinent. His misunderstanding of the Cyprus crisis in 1974 contributed to lasting tension in the eastern Mediterranean that has seriously weakened NATO. His ignorance of the Portuguese revolution that same year had grave consequences in Africa and might have had graver ones in metropolitan Portugal if West German Chancellor Helmut Schmidt had not had the courage to explain to the secretary of state that he did not know what he was talking about.[9] Perhaps most serious of all, for the Persian Gulf and for the United States, was the absurd overcommitment to the Shah of Iran, which led directly to the most humiliating episode in American diplomatic history, the hostage crisis of 1979–81, and to the decline of American influence in a vital region of the world.

The third element of the Nixon-Kissinger plan for extricating the United States from Vietnam involved waging a secret war in Cambodia, complete with brutal B-52 bombing raids. Even more than the first two elements, this committed the administration to a strategy of deception. It became necessary to deceive not only foreign nations and the American public, but also the American media and Congress. It was precisely because the administration relied so heavily on deception for a solution that Nixon's men became so embattled, so obsessed with secrets and spies and disloyalty. Watergate, therefore, was not just the source of a period of weakness in American foreign policy. It was the consequence of a foreign policy brilliantly executed, but poorly conceived and fatally damaged by secrecy.

On Vietnam, which they correctly identified as the key foreign policy issue, Richard Nixon and Henry Kissinger got the worst of both worlds. They did not save South Vietnam from going communist, nor did they succeed in reassuring the world, as Lyndon Johnson had felt it was so important to do, that the United States would stand by its allies to the bitter end. Neither did they succeed in bringing the war to a swift conclusion. It dragged on until 1975, prolonging the agony for the Vietnamese and their neighbors and continuing to leak its poison into the American system.

The Vietnam War and Watergate damaged America's relations with its allies in several ways. To an extent that is unusual in a continent deeply divided by nationality, class, and ideology, the Vietnam War united Europeans against the United States for a time. It is not hard to see why the European left should have been horrified: Those who had been in favor of decolonization found it difficult to accept the American contention that U.S. actions in Southeast Asia had nothing to do with colonialism. The French, after all, had been called colonialists when they had fought the same war over the same ground against the same enemy. The Americans said they had come to protect the Vietnamese, but Europeans knew that familiar refrain; in Algeria and Palestine, in Bengal and Egypt, their ancestors had come to trade or to protect, and had stayed to rule. Indeed, it was the United States, back in the 1940s, that had lectured to Englishmen, Frenchmen, Belgians, and Dutchmen, telling them they had no right and no power to rule the destinies of vast colonial empires. It was not that many Europeans wished to see Southeast Asia experience a communist revolution, or had any illusions about what that would mean. But as Washington and its proconsuls in Saigon made and unmade one government after another, it became harder to square with the anti-imperialist principles that Americans had once avowed and a large majority of Europeans now accepted. To some it looked suspiciously as if the

United States wanted to create an empire, just after the Europeans had let theirs go.

Relations between Europe and the United States worsened steadily through the dozen years after John F. Kennedy's assassination, not only because of Vietnam and Watergate. The mutual irritability between Western Europe and the United States also grew out of Washington's growing indifference to European affairs. Some worried that the American commitment to the defense of Europe might be weakening. The West Germans especially worried about that prospect. At the same time, they and other Europeans resented any American pressure to increase their own defense expenditures.

Not only on security issues were Western Europe and the United States drifting into estrangement. On economic issues the differences were becoming even sharper and more substantial. By 1970 more than fifteen years of rapid economic expansion in Western Europe had utterly transformed the postwar relationship between the two continents. The total level of production in Western Europe was now roughly equal with that in the United States, and the citizens of the wealthier northern European countries were rapidly catching up to Americans in per capita income.[10] Industrial modernization and the rapid growth of productivity in Europe, in fact, hurt U.S. industries in international economic competition.

If the economic growth of Western Europe had been remarkable during the 1950s and 1960s, Japan's was spectacular. When the expansion of the U.S. economy faltered in the late 1960s, the Japanese plowed steadily on, fixing their eyes on long-term targets, seemingly undistracted by the cyclical fluctuations and short-term concern with profit that hypnotized their American and West European competitors. The first significant foreign policy implications of Japan's new economic power impinged on the United States only in the 1970s, when Japanese leaders began to demand consultation on major issues like the American opening to China.

Between 1968 and 1971 the dollar was dethroned from its place as the world's strongest currency, and the Bretton Woods system, under which it had been assumed that the United States would always guarantee the stability of the world's monetary affairs, was virtually dismantled. This development was hailed with no pleasure in European financial centers. The Europeans were proud of their economic progress and of the consequent easing of traditional social tensions, but they felt uncomfortable in the absence of a strong

American lead in monetary and economic matters. The West Germans, in particular, bitterly denounced Washington for "exporting inflation," the result of a devaluation of the dollar.

Relations with the Soviet Union, at first set back by the Vietnam War, improved to a remarkable degree during the first Nixon administration. Henry Kissinger's secret, cynical style of diplomacy went down well with the Soviet leadership, and for a time Moscow was as tempted by the fruits of détente as Washington. Then Watergate altered the Soviet estimate of the balance of advantage. There seemed little to be gained by dealing with a government that could scarcely keep itself in power. Stealthily at first, then more boldly, the Soviet Union shifted from exploring the gains to be won from détente to looking around for neglected trifles that could be snapped up while the great antagonist was distracted and weakened.

However, it would be a mistake to lay too much stress on events in Washington or anywhere else as the basis for sudden tactical shifts in Moscow. Khrushchev had paid for the humiliation of the Cuban missile crisis with his job. The new regime led by Leonid Brezhnev swore to build up its military and economic strength. Heavy sacrifices were made in other sectors in order to equip the Soviet state with a blue-water navy, a massive force of heavy long-range missiles, and conventional superiority in Europe. By the mid-1970s these investments had begun to mature. It was probably inevitable, given the continuity of the Soviet leadership and its unwavering determination to match the United States in global power, that the Soviet Union would resume a more aggressive foreign policy. Adventures in the Horn of Africa, in Angola, in South Yemen, and eventually in Afghanistan were all the more attractive to the Kremlin because after Watergate, and in the disillusion that Americans felt after the withdrawal from Southeast Asia, they could be pursued with little danger of American reaction. To be sure, this Soviet adventurism in turn produced a new and more resolute mood in American foreign policy. But that is another story.

The opening to China, the rise of Japan, the strengthening of Western Europe, the emergence of new, unpredictable regional powers like Brazil and Iran, the relative decline of American economic power, soon to be thrown into sharp relief by the energy crisis—these events were all evidence of a new complexity in a world no longer utterly preoccupied by the superpower relationship between the United States and the Soviet Union. Perhaps the most dangerous consequence of a dozen years of confusion at home, so far as U.S. foreign policy was concerned, was that it had obscured from American eyes the speed with which things were changing elsewhere, thereby postponing the necessary adaptations in the American role in the world.

7 America Among Equals

LESTER C. THUROW

Lester C. Thurow is professor of economics and management at the Massachusetts Institute of Technology. He is the author of several books on public policy and political economy, the best known of which is a study of U.S. economic prospects, *The Zero-Sum Society*. Mr. Thurow has been an economics columnist for the *Los Angeles Times* and *Newsweek* and has written widely on economic issues for a general audience.

THE UNITED STATES entered World War II as one of several nations approximately equal in economic, political, and military power. It left World War II as the world's only economic superpower and one of just two military superpowers. During the first two decades after World War II, America enjoyed both effortless superiority economically and—not coincidentally—diplomatic hegemony among its allies. The latter needed American economic resources to restore their war-torn economies and American military might to prevent Soviet expansion.

America and its allies clashed from time to time over minor concerns, but the United States always prevailed on issues of critical importance to the alliance. As vanquished World War II enemies, West Germany, Japan, and Italy were not in a position to dictate to the victors. France and Britain had to confront the daunting tasks of rebuilding from the devastation of the war and extracting themselves from their rapidly crumbling colonial empires. Japan conceived of its role in the postwar world even more narrowly than the nations of Europe. Interpreting its defeat in World War II as a mandate (and an opportunity) to limit its attention to internal affairs, Japan initially spurned any lasting external political or military involvements outside of its own defense. The Japanese seldom even expressed their opinions on alliance developments during the initial postwar years. In nearly every respect, relations among the postwar allies were asymmetrical: Europe and Japan needed American assistance and cooperation much more than the United States needed theirs. Whenever some of the presumptive "junior partners" did something the United States did not approve of—for example, the British and French invasion of the Suez Canal in 1956—they were promptly disciplined and brought back under control.

By the 1970s, however, the world had become very different. The United States remained one of the world's two military superpowers, but the political and economic factors which had contributed to American postwar hegemony had diminished in strength. As their colonial empires were dismantled, the interests of France and Britain became much more narrowly focused on intra-European affairs. Vietnam had been central to French concerns in the 1950s, but by the mid-1960s few European leaders saw events in Vietnam as a threat to their security or interests. The United States did perceive such a threat. When the British withdrew all of their forces from positions east of the Suez Canal, the United States responded by arming the Shah of Iran to safeguard the Middle East against possible Soviet expansion. Not having staked as much on the Shah's regime, West Europeans did not view his downfall in 1979 as a defeat. Nor did they, later that year, view the Soviet invasion of Afghanistan as a direct challenge to their security or interests. The United States did. In the 1980s the United States also viewed events in Nicaragua with increasing alarm; Europe was less concerned. European and American perceptions, in other words, began to diverge.

Both the Europeans and the Japanese also gradually came to the conclusion that they need not fear a direct Soviet invasion. Soviet intervention occurred elsewhere—in Southeast Asia (Vietnam, Laos, Cambodia), Africa (Angola, Ethiopia), South America (Cuba, Nicaragua), and Southwest Asia (Afghanistan). Although the United States shared the Japanese and European assessment of the unlikelihood of a direct Soviet armed threat to the alliance, it considered Soviet activities elsewhere to be threatening to its own, or the alliance's, strategic interests.

This emerging difference in perception would have made relatively little difference if economic conditions had remained as they were in the first two decades after World War II. If Europe and Japan had remained dependent on American economic support, they would have been compelled to follow America's diplomatic, political, and military lead. Economic conditions did not remain as they were, however. Not only did Europe and Japan recover from wartime devastation, but, less predictably, they also came to challenge America's economic supremacy. Japan became an economic superpower in its own right and several European countries—most prominently, West Germany—began to compete fiercely with American producers on world markets. Europe's collective potential to be a greater economic power than the United States was thwarted only by its failure to act in a coordinated manner.

The nature of international economic relations has also undergone a fundamental change. Despite the volume of postwar trade, a truly integrated world economy did not come into existence until very recently. This world economy emerged partly as a result of an increase in the volume of trade, which significantly affected the level of international economic integration, and partly as a result of a shift in the prevailing patterns of trade. Prior to World War II, nations with advanced economies protected domestic producers from international competition through the use of import tariffs and quotas, while conducting much of their trade with captive colonial or neocolonial markets. After World War II, trade was increasingly conducted between advanced industrial economies.

Prior to World War II, few nations' economies could be adversely affected by shifts in the volume or pattern of international trade or changes in the economic status of other countries. National economies were primary; international trade and the world's other economies were secondary. By the 1980s trends in the world economy and developments within other national economies were capable of significantly influencing the economy of practically every nation in the industrial world.

When the new Mitterrand administration tried in 1981 to stimulate the economy of France with old-fashioned Keynesian policies, the same policies that had worked for the American administrations of Kennedy and Johnson two decades earlier, it found that it could not achieve its goal. France was simply too open to imports from the rest of the world and the desired national expansion sank in the quicksands of mounting trade deficits. Within Europe the slow recovery from the 1981–82 recession is attributed in large part to high American interest rates and vast American budget deficits. Europeans sense a crucial loss of control over their domestic economies.

At the same time, never before has a military power led an alliance in which the member states were capable of challenging its economic hegemony. All of the nations in the Western alliance are major players within the emerging world economy. This has led to the peculiar situation in which America's political and military allies are at the same time its economic competitors—and the competition is taken seriously by both the United States and its allies. A successful Japanese auto industry raises employment and income in Japan, but lowers income and employment in the United States. Nagoya prospers while Detroit freezes. Economic competitors have historically become military competitors. But, ironically,

America's greatest military competitor, the Soviet Union, cannot threaten it economically, while its major economic competitors pose no military challenge.

The tasks of leading such an unusual alliance are quite formidable. For the alliance to succeed, it must develop some technique for controlling the major adverse effects of international competition without slowing the overall expansion of the world economy. If economic competition is not to undermine the political cooperation upon which its collective military strength depends, the alliance will require a degree of economic leadership as advanced as its current military leadership.

If America's allies had simply caught up with its economy and all were moving ahead at a relatively equal pace, the tasks of managing the new world economy would be less difficult. The level of economic management demanded by a world economy becomes possible only if one of its members has an economy sufficiently strong that it can advocate policies (e.g., trading rules and regulations, monetary and fiscal policies, coordination of exchange rates) which are in the general interest, even when they are not necessarily advantageous to its own economy. If the prospective leader's economy is so weak that it cannot set aside its own direct economic self-interests, it cannot assume the general management role that is required. Moreover, it cannot, by the power of its example, persuade other nations that they ought to set their own narrow self-interests aside.

One could argue, for example, that the Great Depression was partly caused by Britain's abdication of its world financial leadership. No other country was at the time willing or able to step forward and take up the financial burdens the British had laid down. The Depression began in the 1920s with a multinational series of financial crises, and defensive measures were never organized to prevent their spread. Financial crises were not new. Many had occurred before, but British financial might had always held them in check. By the 1920s Britain's own economic weakness rendered it incapable of playing its traditional role.

To work as it is now constituted the world economy must be managed. Enormous quantities of money move around the globe very rapidly and sudden shifts have the potential to harm any economy. The volume of trade is so large that unexpected changes in trade flows can threaten the jobs of millions of workers and the existence of entire national industries. Acting alone, individual governments can safeguard their economies only by putting in place protectionist measures that undermine the international capital and goods markets that are now emerging. The national response is thus to retreat into an economic fortress. Unfortunately, this process is well under way in many countries.

While America may not want to play the required management role—and will not be able to do so if it becomes too weak economically—it has no choice but to try, if it wants to maintain the military alliance of democratic countries. There is currently no other country that can step in and fill the management role if the United States abdicates. Admittedly, it will no doubt prove frustrating for the United States to play a managerial role on a board of directors—comprised of the governments of its allies—where previously it had acted as if it were a sole proprietor.

America thus faces an economic problem that is simply put. The huge technological edge enjoyed by Americans in the 1950s and 1960s has disappeared. The country's effortless economic superiority has been overcome by competitors who have matched its technological achievements and may soon surpass them. Given present productivity trajectories, the per capita output of the American economy may shortly fall below those of its allies.

To complicate matters further, at the same time that the United States has lost its effortless economic superiority, its economy has been absorbed into a world economy. There now exists a world market for the exchange of most of the goods and services produced in the United States, not just an American market. Competition is worldwide, not just American. The United States faces the novel task of learning how to compete in the new world economy at a time when its relative economic strength is less than it has been at any moment since World War II.

This message was transmitted to Americans in a series of body blows seemingly punched from abroad. The economic strain of the Vietnam War, foreign crop failures that resulted in rising domestic food prices, two OPEC-led oil price shocks, industries wounded by foreign competition retreating behind government protection—all of these developments have, not surprisingly, contributed to a new feeling of economic vulnerability and estrangement among Americans, a sense of having been betrayed by an economic order they were instrumental in creating.

Productivity—output per hour of work—is the best general measure of a country's ability to generate a high and rising standard of living for each of its citizens. The standard of living enjoyed by the citizens of any nation is at all times limited by what they are able to produce. To consume more, a country must produce more.

Productivity also determines a country's ability to maintain high relative wages in world market competition. Under conditions of declining productivity, a nation can only sell its products if its wages decline correspondingly. American workers have been less inclined than most to accept such an outcome. Americans would like to compete from a position of relative wage superiority, though they will probably have to learn to accept a position of wage equality; they certainly do not want to compete from a position of inferiority.

Manufacturing is probably the best place to compare American productivity to that of its competitors. Since all manufactured goods are potentially tradable and essentially similar from country to country, the measurement problems which plague the international comparison of living standards are much less formidable. The evaluation of manufacturing data does not entail value judgments.

The data in Table 1 show the 1983 level of manufacturing productivity and recent productivity growth rates for seven leading industrial countries. As the table indicates, the absolute level of U.S. manufacturing productivity has already been surpassed in Germany and France. Since many of the small northern European countries not appearing on this table have achieved productivity levels similar to that of Germany or France, much of northern Europe (with the exception of Ireland and the United Kingdom)

TABLE 1. Comparative Manufacturing Productivity

Country	Output per hour of work[1] 1983	Rate of growth[2] 1977–82	1983
United States	$18.21	0.6%	4.2%
Germany	20.22	2.1	4.6
France	19.80	3.0	6.1
Italy	17.72	3.6	0.6
Japan	17.61	3.4	6.2
Canada	17.03	−0.3	6.9
United Kingdom	11.34	2.7	6.1

may now be slightly ahead of the United States in manufacturing output. As the data for Italy demonstrate, parts of southern Europe are not far behind.

Many are surprised that Japanese productivity still lags behind that of the United States. This can be attributed to the fact that the Japanese manufacturing economy contains a peculiar mix of superefficient and poorly performing industries. Japan's inefficient industries do not export, however, rendering them invisible to American eyes. Yet even though overall manufacturing productivity is slightly lower in Japan than in the United States, it ought to be remembered that it is possible to drown in a river which is "on average" two feet deep. In crucial exporting industries such as steel, automobiles, and consumer electronics, Japanese manufacturers are second to none.

While America's economic superiority has clearly dissipated, its current position is probably best described as one of parity, not inferiority. The rest of the advanced industrial world has essentially caught up with the United States. According to the data, Germany's output is approximately 10 percent higher than America's, but given the vagaries of such measurements, this does not represent an unambiguous lead.

If one examines comparative rates of recent productivity growth, however, it becomes clear that the American economy could soon occupy an inferior position among the world's leading economies. America's rate of productivity growth has been below those of Europe and Japan since World War II, but now that the latter have essentially matched U.S. manufacturing output, their continued higher rates of output growth pose a major economic challenge. From 1977 to 1982 American manufacturing productivity grew at the rate of 0.6 percent per year, one-third the rate in Germany (2.1 percent), one-fifth the rate in France (3.0 percent), and approximately one-sixth the rate in Italy (3.6 percent) and Japan (3.4 percent).

Furthermore, manufacturing productivity remains one of the bright spots in the U.S. economy, having grown three times as fast as productivity in all private industry during the 1977–82 period.[3] The rate of private sector productivity growth has undergone a steady decline in the postwar period, from 3.3 percent (1947–65), to 2.4 percent (1965–72), 1.6 percent (1972–77), and most recently 0.2 percent (1977–82).[4]

In 1983, as its economy recovered from recession, the United

States experienced a cyclical upturn in nonfarm business productivity (3.1 percent).[5] However, this did not reflect a fundamental change in overall productivity trends. Productivity always rises rapidly with the onset of a recovery, because overhead costs are distributed over a higher quantity of output. In fact, 1983 productivity gains were much weaker than what could have been expected if the United States still had the advantage of its 1947–65 productivity trends. If this were the case, the 1983 cyclical upswing in productivity growth would have been at least twice its actual size.

Despite the fact that the United States was experiencing a stronger cyclical recovery than any of its allies in 1983, its manufacturing productivity growth still remained below that of its competitors. Thus, in one of its better years American productivity growth was still surpassed by the rates attained by Europe and Japan in what was for them an unsatisfactory year. And while it is too early to determine whether there will be any further cyclical gains from the recent recovery, U.S. productivity grew very little in the last half of 1984. The downward secular trend in American productivity growth has persisted through four such recoveries and seems likely to last through a fifth.

To remain a world-class economy, American productivity must be approximately equal to that of the best industrial economies. Though the United States has not yet lost its world-class status, its rate of productivity growth must be accelerated if it is not to do so. The present sense of economic vulnerability in the United States is grounded in a reality that may worsen.

In the 1950s and 1960s American firms ventured abroad and established multinational enterprises. In the 1970s and 1980s foreign firms reciprocated, building U.S. subsidiaries for the service of the American market and establishing other production facilities for their international trading activities. Some American firms have become subsidiaries of foreign firms as well. The domestic market for American automobiles has dissolved in a flood of imports and foreign-owned assembly plants. France's Renault purchased American Motors. Volkswagen of Germany and Honda and Nissan of Japan have all opened plants in the United States. The auto market has been internationalized.

As a proportion of the U.S. gross national product (GNP), exports more than doubled between 1970 and 1981, from 6 percent to 12½ percent.[6] Although by 1984 exports had fallen as a consequence of the dollar's high value overseas, imports were still commanding 12½ percent of the American market.[7] The significance of this fact is revealed when it is recalled that Japan exported only 16½ percent of its GNP in 1981.[8] The United States is now approaching Japan's level of dependence on international trade. If one excludes intra-European trade, Europe's exports, as a proportion of its aggregate GNP, are not far above America's. The key difference is that the Japanese and Europeans recognize their dependence on international trade and the United States does not.

Essentially, Americans live in a world in which everything that can be traded is, or eventually will be, traded. With a little ingenuity, products and services previously considered immobile have recently become tradable. The assembly of buildings from prefabricated components is an example of this development. Firms that do not work to capture foreign markets have found that they must spend excessive resources defending their home markets. Conversely, firms with secure home bases have a competitive advantage on foreign markets.

Threatening international competition exists not just in old-line "rust belt" industries, such as steel or autos. The challenge is just as strong at the front end of the U.S. economy, in new products and new production technologies—the so-called Silicon Valley in northern California or the industries along Route 128 near Boston. Consider home video recorders, a very recent product made with new production technologies. Although some of these are marketed under American brand names, not one has ever been made in the United States; they are 100 percent imported. Semiconductor chips, the technological core of much of today's computers and electronics, were once an American monopoly. But by the end of 1983 the Japanese had captured 70 percent of the vast market for 64K RAM chips, and by the end of 1984 held an even larger share of the newer market for the higher-capacity 256K RAM chips.[9] At the end of 1982, for every ten thousand employees Sweden had thirty production or assembly robots, Japan had thirteen, and West Germany had five. The United States had four.[10] The robots are being used not to make older, well-developed products that typically drift off to lower-wage countries under the effects of the product cycle and comparative advantage. They are being used for new products, an area in which the United States competes with high-wage allies that have begun to joust for world economic leadership.

The world market is a reality that most Americans are reluctant to accept. No one likes to pass from a secure position of economic superiority to one of insecure, competitive inequality or even inferiority. In devising an appropriate policy response to this situation, the United States has two divergent choices: It can do what is necessary to become more internationally competitive while working to preserve the increasing integration of the world economy, or it can seek to withdraw from the international economy, pursuing an economic-fortress strategy in relative isolation.

To do the latter would contribute to the disintegration of the world economy and to the disruption of current political and military alliances. An American economic withdrawal would force Europe and Japan to make major changes in the ways they administer their economies. Without open access to the American market, for example, many Japanese industries would suddenly be grossly overbuilt. The economic-fortress strategy would be received abroad as a declaration of international economic war.

Within the United States, a fortress strategy would mean that living standards would grow more slowly than they had in the past, perhaps gradually falling behind those of the rest of the alliance. If it were not for envy, such a decline might be politically manageable. The Soviets have achieved rough military parity with a per capita GNP that is half that of the United States. Americans, however, are not inclined voluntarily to spend a growing fraction of their income defending an alliance that is composed of nations which are wealthier than the United States.

By the mid-1980s Americans had yet to move decisively on either of these two strategies for dealing with economic vulnerability. Many economic policymakers still apparently hope that they will not be forced to choose. The present tendency, however, is unmistakably toward protectionism and the fortress strategy. In industry after industry around the world—from textiles to autos to machine tools—"managed trade," in which governments seek to regulate export and import flows, has been substituted for what had previously been a free-trading regime. As other members of the alliance follow suit, the world economy is bound to contract and become less integrated.

There is substantial evidence that this kind of contraction is already under way. The fraction of the world's GNP comprised of exports appears to be falling. No one currently knows whether this is merely a temporary anomaly in the trend of the past thirty-five years toward increasing economic integration or the onset of a long-term slide toward disintegration. Normally the leader of a political and military alliance would be expected to stand at the forefront of efforts to resist such disintegration, but because of the relative weakness of its economy (nowhere more evident than in the $123 billion trade deficit in 1984) the United States has shaped up as a leader in the gradual shift toward protection and less economic integration.

It is possible to argue that increasing protection is a realistic response to present conditions. The current level of economic integration does not seem viable, given the failings of the American economy and the inability or unwillingness of alliance governments to coordinate their monetary and fiscal policies, as demonstrated by the repeated failures of economic summits. It may be wise, some maintain, to minimize potential political damage to the alliance by applying protectionist measures in an effort to stifle divisive economic competition. However, the increasing incidence of protection is not coming about through a rational policymaking process. Even if less integration is to be the objective, unplanned movement toward protectionism is far more dangerous than a multilateral decision to reduce economic integration. A drift toward protectionism is apt to get out of hand and become a flight. When several nations seek to protect their industries at once, everyone's exports fall, economies contract, and there arise increasing pressures for even more protection. It was precisely this self-perpetuating cycle which worsened the impact of the Great Depression during the 1930s.

In 1984 the United States began to recover from its ninth recession since World War II. The four it has experienced since 1961 were all deliberately induced by the federal government as a means of controlling inflation. The causes of the inflation differed in each instance, but all of the triggering events appeared to originate from abroad. The fact that American foreign policy has been part of the cause each time does not make it any less disconcerting for Americans; indeed, they are troubled by the discovery that they have an economy which can apparently be dominated by "foreign" events.

Whatever its cause, once inflation has broken out, the U.S. government has had the benefit of a single cure: to slow the pace of the economy with tight monetary and fiscal policies, deliberately increasing unemployment and squeezing inflation out of the system. During 1984 and 1985 the United States enjoyed an economic upturn, but the parameters of the system had not changed. The next inflationary shock—likely to occur when the overvalued dollar falls relative to other currencies—will require the application of the same recessionary cure, and once again Americans will perceive the inflation as an international typhoon buffeting their tranquil domestic economy.

The recessionary cure is also becoming more painful. The depth and duration of the recessions necessary to control inflation seem to be growing. Yet it is difficult to manage a healthy capitalist economy if it must spend longer and longer periods with its engines in reverse. The industrial structure weakens financially as the government is forced to provide increasing amounts of aid and protection to the industries that have been most harmed by recession. The tendency of the industrial structure to be undermined is becoming visible both in the United States and abroad. The unprecedented number of bank failures in the United States is perhaps the most dramatic recent example. Sometimes even relatively trivial events, such as the temporary shutdown of seventy-one Ohio savings and loan associations, can seem to have an impact on important variables like the value of the dollar.

As world markets, particularly those for capital, are further integrated, rising American interest rates are more likely to cause higher interest rates worldwide, thereby slowing not just the U.S. economy but the economies of many other nations as well. America's alliance partners assert that its "stop-and-go" policies and high interest rates contribute significantly to their domestic economic difficulties. In Europe the fundamental economic problem is jobs. Although its "baby boom" generation is just now beginning to enter the labor force, Europe, in the aggregate, has not generated any new jobs since the early 1970s. Japan faces the fundamental dilemma of possessing an economy that cannot grow unless its exports expand faster than domestic sales. Such a situation cannot continue indefinitely, however, because the rest of the alliance will not tolerate the unrestrained expansion of competitive Japanese exports.

Each of these economic predicaments creates irresistible pressures for import restrictions. Spreading protectionism threatens the political and economic ties which the United States has helped to forge since World War II. Realistically, political alliances cannot remain fully intact under conditions of economic warfare.

Quite apart from the erratic performance of the U.S. economy during the past several years, the Reagan administration has fundamentally abdicated America's role as an economic leader of the alliance, because of its ideological commitment to the idea that economies do not need to be managed. According to this perspective, capitalist economies of any scale are self-correcting when left alone, but are inefficient when managed administratively. Nowhere is this more evident than in the Reagan administration's reluctance to intervene in foreign currency markets to prevent the highly valued dollar from chasing American agricultural and industrial products out of international markets. If the administration will not intervene in a major way to protect its own producers, then it certainly is not going to intervene in the world economy in a manner which protects the economies of other countries.

Given the eventual political costs of international economic drift, it is worth considering a number of objections to this free-market view. First, the realization of the self-correcting properties of a free-market economy demands that no actor interfere with the functioning of the market. This condition has never been met, either in the United States or abroad. The United States has at times protected its industries as conscientiously as any of its allies, and for many of the latter, government intervention is an essential feature of economic life. Neither the United States nor its allies are about to abandon such practices entirely.

Those who have faith in the self-managing qualities of free markets also tend to ignore the ever-present need for governments to create recessions to stop inflation. Regardless of the equity of this strategy, it works much less successfully when it is not pursued in a coordinated fashion; it often occurs that one part of the alliance is seeking to expand through easy monetary and fiscal policies while another part tries to contract with tight monetary and fiscal policies. When the United States in particular decides to put on the economic brakes, as it did between 1979 and 1982, the contraction of its economy is unwittingly imposed on the rest of the alliance through open capital markets and trade flows; foreign interest rates rise and foreign exports fall. In 1981–82, for example, the planned American contraction completely overwhelmed France's attempted expansion.

The purist free-market view also neglects the fact that markets generate losers as well as winners. This applies to entire national economies as well as specific industries. Americans are instinctively attracted to the free-market ideal because they assume that the

United States will be a winner. This is not always the case, and when it is not, Americans quickly abandon their free-market principles.

Finally, implicit in this perspective is the mistaken assumption that politics and economics are distinct and isolated realms. In the equation between economic policies and political policies the arrows of causation, like those in physical chemistry, point in both directions.

Some of the consequences of an international economic policy of benign neglect can be seen in the problematic interactions of the world economy and the American middle class. The United States has for some time considered itself a middle-class country with a political system that is uniquely suited to—and dependent on—its particular social and economic structure. Yet under the pressure of international competition, the American middle class is becoming an endangered species. If a middle-class household is defined as one whose income falls between 75 percent and 125 percent of the national median income (a range that extended from $15,664 to $26,106 in 1983), then the American middle class has declined from 27.1 percent of all households in 1968 to 23.2 percent of all households in 1983 (see Table 2). Varying definitions of what constitutes a middle-class income affect estimations of the size of the middle class, but they do not refute the existence of this trend.

Similarly, if a middle-income job is defined as one that pays between 75 percent and 125 percent of the median annual salary, then the number of middle-income jobs traditionally held by males actually dropped from 13.6 to 13.3 million in 1983, despite the creation of millions of new jobs during that year's recovery.[11] The middle class is disappearing: Roughly one-third of those families that were formerly part of the middle class have fallen below the

TABLE 2. The Changing Middle Class[12]

Year	Bottom (0.0 to 75% of median household income)	Middle (75 to 125% of median household income)	Top (125% and up of median household income)
1968	36.2%	27.1%	36.7%
1983	37.5	23.2	39.3

designated income range, and approximately two-thirds of these have risen above it. A bipolar distribution of income is gradually replacing the one dominated by middle-income levels.

Several factors are causing this decline of the middle class. Most middle-income households become low-income households when the prime earner is unemployed, as occurred quite frequently in the late 1970s and early 1980s. Female-headed households, which more commonly fall in the low-income range than male-headed or dual-income households, comprise a rising proportion of the total. The integrity of the middle class is also being affected by the special characteristics of America's new growth industries. The microelectronics industry, for example, tends to employ large numbers of low-wage assemblers and high-wage engineers and designers, but provides relatively few middle-income jobs. This results in part from the nature of the technology and in part from its nonunionized environment. In several other industries, where unions had previously succeeded in obtaining middle-income pay scales for relatively low-skilled jobs, the unions themselves have been experiencing a period of rapid decline.

Perhaps the most significant factor in the decline of the middle class has been the beating that American industry is taking in international trade. The U.S. middle class earns its living in many of the most internationalized sectors of the economy. Industries such as automobile, steel, and machine tool manufacturing, which have traditionally provided millions of American middle-class jobs, are precisely those which have fared worst in foreign competition. When these industries shrink, the middle class shrinks with them. At current GNP and productivity levels, the $123 billion the United States spent on foreign imports in excess of its own exports could have provided for three million largely middle-class jobs.

The implications of these trends are potentially grave. According to the conventional wisdom, a strong middle class is essential to a healthy liberal democracy. Societies characterized by income polarization are commonly thought to be prone to instability. Karl Marx foresaw the potential for revolutionary upheaval as societies became progressively polarized between the capitalist rich and the proletarian poor. Marx did not, however, foresee the ability of capitalism to generate a class that would devote itself to the preservation of that system, would work to alleviate the system's worst excesses through social welfare programs, and would represent for the poor the promise of someday escaping from their poverty. Not surprisingly, the revolutions Marx anticipated have occurred in essentially feudal societies like Russia and China, where economic development had yet to produce a middle class.

If the conventional wisdom is correct, and a vital middle class truly serves as a source of social cohesion, then the United States may be in the process of becoming unglued as a consequence of economic competition from its allies. A democratic superpower requires domestic consensus, tranquility, and economic success. If the middle class perceives the competition of U.S. allies as a threat, its support for the objectives of the alliance will falter. The Japanese, for example, may increasingly come to be viewed as free riders on the political and economic systems of others, exporting their unemployment to the rest of the world in the form of trade surpluses, and failing to shoulder a fair share of the alliance's military costs. These sentiments have thus far been restrained in the United States, but they will eventually become more widely held if the middle class continues to shrink.

Many people believe that the United States could remain a world-class military power without having a world-class economy. If the 41 percent of the U.S. GNP spent on defense in 1943 is contrasted with the 6 percent of the GNP spent on defense in 1983, and if one considers that real per capita disposable personal income was twice as high in 1983 as it was in 1943, it becomes clear that a gradual economic decline relative to the rest of the alliance would hardly deprive the United States of the funds required for military preeminence.[13] To remain a superpower, the United States must simply be willing to spend what is necessary, but what is presently necessary does not significantly affect the life-styles of American citizens. The military burdens that Americans currently shoulder are quite modest relative to what they have. In other words, in this view economic success and military power seem no longer as closely coupled as they once were.

In practice, however, the facts are otherwise in a democracy. Substantial sacrifices will be made in the face of an immediate external threat, but even very modest sacrifices in personal living standards are unlikely to be tolerated for a long-term goal like the exercise of "world leadership." This is particularly the case if the leader's allies seem to be threatening its economic position or failing to assume a fair share of the joint military burden. With its command economy and authoritarian political system, the Soviet Union can maintain its superpower status without a world-class economy. The United States cannot.

Americans have yet to confront the reality that remaining a military superpower makes it more difficult to be an economic superpower. Military spending is a form of consumption; on balance, it does not enhance a country's ability to produce more goods and services in the future. As a consumption expenditure, high military spending will not be a drain on the U.S. economy if the revenues needed to pay for it are raised from taxes on other forms of consumption. When military consumption goes up, other forms of consumption will simply go down.

However, if military spending is financed in part by decreased expenditure on investment in new plants, new equipment, or in the labor force, then that spending does become a hindrance to future economic success. In a democracy, it is perhaps inevitable that at least some defense spending will come at the expense of investment, for it is always politically easier to spend less on activities that do not affect current living standards. Yet, if the United States were to reduce the fraction of its GNP currently spent on defense to the Japanese level, and could transfer these funds to plant and equipment, it could close the investment gap between itself and Japan.

While the problem of financing a large military budget is, in principle, solvable through increased taxation of private consumption, there is an associated human problem that is less easily resolved. Defense-related research and development (R&D) is more attractive than commercial R&D, since it is typically at the technological cutting edge and is subject to few of the budget constraints that affect its commercial counterpart. If a "Star Wars" defense against nuclear attack could be built, there would be tremendous pressure to do so, regardless of its cost. In the civilian economy, new products must be more cost-effective than their predecessors. As a result, the best and the brightest among America's engineering and science prospects tend to enter military R&D. It is simply more fun. The United States consequently leads the world in the development of new military hardware, but many of its competitors have begun to overtake it in commercial innovation. In this manner, the large U.S. military establishment handicaps future civilian economic success.

All told, America's economic problems are now intertwined with world economic problems in a manner different from anything the United States has experienced in the past. Not surprisingly, the result is an unsettling feeling. Americans do not want to slow economic progress in the rest of the world. After all, it is better to live in a rich world than in a poor one, even if one sometimes envies one's neighbors. Not only are wealthy neighbors apt to be more agreeable than poor ones, but they are also certain to provide better markets for the goods and services Americans would like to sell. While catching up with the U.S. economy, several countries have grown faster than the United States for extended periods of time. But there will inevitably come a time when the United States will want to accelerate its economic performance to keep from falling behind the rest of the industrial world.

The United States can never return to the era of effortless superiority, but it can rebuild its industrial structure and remain an equal among peers. In a sense, the United States may need to experience the moral equivalent of defeat. No society will undertake the painful tasks necessary for economic reconstruction unless it understands that it has been defeated—that it must rebuild in new ways or lose a valued economic position. New institutions have to be created, new processes have to be implemented, and new habits have to be formed.

The sense of defeat was the advantage enjoyed by Japan and Germany after World War II.[14] It was not so much that their factories were destroyed, for that also happened in Great Britain. The real advantage of defeat was social: the willingness to change old ways for new ones, to reject imperial pretensions.

Japan's renowned system of cooperative labor-management relations, for example, was not an ancient historical inheritance, as is widely believed, but something that was carefully planned in the 1950s. Even today it is missing in some sectors, as revealed by the virulence of labor-management relations on the Japanese national railways. To assert that defeat is a necessary motivational force for change is not to assert that it is a force which Americans enjoy experiencing. Americans have, indeed, begun to sense an economic defeat, but this has not yet had the socially jarring impact of a military defeat.

For the United States to become competitive with the rest of the world, it will need to make progress on many fronts simultaneously. Wherever key inputs into the American economy do not match the quality of the competition, those inputs will need to be brought up to world-class levels. America can hardly compete when its rate of capital investment (as a proportion of GNP) is half that of the Japanese and two-thirds that of Europe.[15] Nor can it keep pace with its competitors while perenially investing less in civilian research and development.[16] When only two of the ten most highly rated cars sold in the United States, as ranked in consumer surveys, are designed and built domestically (Lincoln and Mercury), when the mathematical ability of its labor force is half that of the Japanese, and when its basic industries must cope with a repeating cycle of recession, high interest rates, and an overvalued dollar, the American economy clearly lacks the resilience and vigor of its competitors.[17] If American interest rates are three times those of Japan, as they were in 1981–82, then Japanese firms have a major cost advantage even before American production lines start moving.[18]

Instead of rising to the occasion, it will doubtless prove easier for Americans to blame their economic problems on their competitors. These sentiments have already been expressed in corporate advertisements demanding a "fair" or "level" playing field. But what Americans perceive as a level playing field will surely be seen by others as one on which America inevitably wins. The rest of the world will consider American demands for favorable changes in economic rules or practices to be self-serving—and rightly so. For if the major capital, labor, and management inputs into the American economy are not of world-class quality (and they are not, at present), then the United States will only be able to compete with a set of rules that is stacked in its favor.

In the recent cyclical boom, many Americans were persuaded that the country could return to a position of unchallenged preeminence. Trade deficits can be ignored, as they have been, for short periods of time. But no country can continue borrowing sufficient funds to finance a $123 billion, and growing, annual trade deficit. At some point, though no one knows precisely when, foreign investors will decide that they have enough of their funds at risk in the United States and will stop lending. The dollar's value will fall as foreigners stop buying it, while Americans attempt to purchase foreign currencies to finance their trade deficit. As a consequence, the price of imports and import-competing goods will rise, leading to another external inflationary shock. Monetary authorities will fight the new inflation by slowing the growth of the money supply and raising interest rates, and the economy will probably enter its tenth postwar recession. Cyclical euphoria will once again be undermined by structural reality.

The military viability of the alliance ultimately demands that the United States achieve a competitive rate of productivity growth and take the lead in coordinating international monetary and fiscal policies. If either of these tasks proves impossible, the long-term security of the alliance cannot be assured. Although current political and military ties can endure without American economic dominance, they cannot persist under circumstances of American economic inferiority.

A movement toward managed trade and away from economic integration will be the most likely outcome in the years ahead. Such a movement would not be an ideal outcome, since only successful integration can lead to the highest possible standards of living. But even economic semi-isolation is to be preferred to a policy of benign neglect, for the latter will find the world drifting toward protection in a manner that is liable to turn into a panic during the next recession.

8 Uncle Sam's Hearing Aid

ALI A. MAZRUI

Ali A. Mazrui is research professor at the University of Jos in Nigeria and professor of political science and Afroamerican and African studies at the University of Michigan. Former president of the African Studies Association, he has written widely on the philosophy, politics, and international relations of the Third World. His books include *Towards a Pax Africana: A Study of Ideology and Ambition, Africa's International Relations: The Diplomacy of Dependency and Change,* and, with Michael Tidy, *Nationalism and the New States in Africa.*

AMERICANS ARE brilliant communicators but bad listeners. Because Americans can communicate effectively, the rest of humanity is, to some extent, becoming Americanized. But because Americans are bad listeners, they have resisted being humanized, in the sense of learning to respond to the needs and desires of the rest of the world. The Iranian revolution, which led to the overthrow of the Shah in 1979, was an excellent illustration of this one-way traffic. For years American culture had been transmitted through amplifiers and loudspeakers to Iran, while Uncle Sam switched off his own hearing aid and turned a deaf ear to the return messages being transmitted by a resurgent Islam. Iran was becoming Americanized and Westernized, but under protest. The United States did not hear the protests until it was too late. This problem, and many others like it, go a long way toward explaining the growing distance between America and the Third World.

The means of communication at the disposal of the United States have to be distinguished from actual messages it transmits. Americans use a number of different languages in communicating with the people of the Third World. One is the language of production. Because its economy is the largest in the world, the United States can use its productive power as a medium of protest or disapprobation. That is what Jimmy Carter did when he imposed an embargo on grain sales to the Soviet Union after the Soviet invasion of Afghanistan. But the moral message of such an embargo was easily neutralized four years later when the United States invaded Grenada and stepped up is military activity in Central America. The United States effectively communicated its own message of disapproval to the Soviet Union for harassing a small neighbor, but it seemed puzzlingly unaware of the moral implications of its own similar behavior.

Another medium available to the United States for international communication is the language of the consumer. Because the United States is the largest market in the world for a wide range of goods, it has considerable consumer power. When Idi Amin was ruling Uganda, the United States was sufficiently aroused morally to impose an embargo on the purchase of that nation's coffee, a third of which was, at the time, consumed in the United States. Yet when international sanctions were imposed on the white-minority government of Rhodesia, the United States violated them, and it continues to argue against any serious economic sanctions against the white regime that rules South Africa.

A third form of language available to the United States is that of currency or liquidity, as distinct from commodities. The United States has immense power over international financial institutions like the World Bank, the International Monetary Fund, and, more generally, the commercial banks of the Western world. Because the Reagan administration has not listened to the groans of the world's poor, the United States reduced its contribution to the budget of the World Bank's International Development Association (IDA) in 1984, and other nations followed suit. The IDA provides development loans on favorable terms to the poorest of the poor countries. Using its power to reduce the IDA's effectiveness is an instance of singular American insensitivity. Similarly, the United States withdrew from the United Nations Educational Social and Cultural Organization (UNESCO), having decided that it does not want to listen to some of the messages emanating from UNESCO and other U.N. agencies, even if those messages have overwhelming majority support. U.S. withdrawal is a way to switch off outside comment, as well as an exercise of America's purse power. On the other hand, by supporting a large IMF loan to South Africa, the United States used its financial influence to transmit an ideological message of a different sort, which again demonstrated indifference to the moral concerns of the rest of the world.

Another language the United States employs is that of technology. It has been used more often against the Second World (the advanced communist countries) than against the Third World—most notably, the embargoes on the export of advanced technology that followed the imposition of martial law in Poland and the Soviet invasion of Afghanistan. The Reagan administration also attempted to obstruct the development of a Soviet natural gas pipeline to Western Europe by withholding key materials and technical data. Toward the Third World, the United States has imposed an embargo on the export of nuclear technology. America has simply decided not to communicate with the Third World in this area, establishing a conspiracy of nuclear silence, which is itself a loud vote of no confidence.

In the face of this nuclear secrecy one presumptive American friend, Pakistan, appears to have embarked on a strategy of industrial espionage. Implicit in its effort to obtain fission technology is the conviction that the world cannot be divided between nuclear Brahmins and non-nuclear untouchables. If nuclear disarmament is not universal, it will not be tenable for long. It is not so much that the United States has failed to listen to the grievances of the nuclear have-nots, but that it does not even seem to be aware that the present nuclear caste system is intrinsically unstable. Nuclear proliferation cannot be stopped until the superpowers disarm.

Weapons power is yet another language available to the United States. Central America—particularly Nicaragua—is the latest arena in which the flexing of American military muscles serves as a mode of militant communication. Anti-Sovietism and anti-Castroism are the substance of these recent messages, but the implicit threat is that if American warnings are not heeded and Marxism is not banished from the Western Hemisphere, severe consequences will follow. In many instances before Central America —from Nasser's Egypt to Vietnam to Mengitsu's Ethiopia—Third World radicals have had to contend with American pressures, and sometimes even conspiracies, against them.

A final language available to the United States is, quite simply, the English language, the most widely understood tongue in history. English does not, of course, have the largest number of speakers in the world—that distinction belongs to the Chinese language—but English has been adopted as the primary medium of national business in more countries than any other language, and it also has the largest number of non-native speakers. It is meaningful to speak of America's linguistic export. While it was Great Britain which initially carried English to so many of the world's countries, it is now mainly the United States which determines how many individuals learn English worldwide. It is America's global power that gives prestige to present-day English. Unfortunately, the United States is better at using English for transmitting its messages than it is at using the language to listen to the whispers of the rest of humanity.

It is difficult to disentangle American means of communication from their content, but one should at least attempt to be more explicit about some of the messages that the United States has been communicating to the Third World.

Americans often confuse the two, but, in fact, capitalism and democracy have in recent years asserted competing claims on the content of America's international communications. Capitalism may be thought of as the doctrine of competitive economics, resulting in market forces. Liberal democracy may be thought of as the doctrine of competitive politics, resulting in political pluralism. On balance, the United States has been much more successful in transmitting capitalism than in transmitting democracy.

The Carter administration's emphasis on human rights in its foreign policy sprang from a desire to export liberal democracy. In contrast, the Reagan administration has emphasized the export of capitalism, which explains the enhanced role of private enterprise and free-market ideology in that administration's international economic and political relations. An excellent illustration of this phenomenon is the so-called Economic Policy Initiative, in which African nations were promised extra American aid if they would reform their economic systems along capitalist lines. Of course, the distinction between Carter and Reagan is not quite so sharp in practice as in rhetoric. Carter's human rights policy was inconsistent in places like South Korea and the Shah's Iran, and Reagan's anticommunism does include a concern for human rights. Yet the difference in emphasis between the two administrations still has practical significance.

If the United States were consistently to emphasize the sanctity of human rights, that would probably be good news for the rest of humanity. It would mean that U.S. foreign policy was beginning to be humanized. If the United States were to stress the sanctity of the profit motive, however, that would probably be bad news for humanity, for it would imply the Americanization of the human race rather than the humanization of America. This is not because the profit motive is necessarily evil in itself. Capitalism has served the United States and Western Europe very well; even the People's Republic of China has gingerly begun experimenting with market forms. But while a U.S. crusade for human rights could be genuinely moral and altruistic, a crusade for capitalism would be suspect among the developing nations. They know that the United States is the heartland of the current international capitalist system, and therefore its chief beneficiary, and so they fear that Americans would be disinclined to restructure that system in the interests of others or to put at risk their own privileged position.

When the United States has been in a phase of genuinely trying to promote human rights, it has reduced its support for repressive regimes in the Third World, as when Jimmy Carter cut economic and military assistance to Guatemala and Uruguay because of abuses in those countries. But when the United States has sought to consolidate capitalism—as in Mobutu Sese Seko's Zaire—there has seldom been room in its calculations for elaborate notions of democracy and social justice. America's hearing aid has been more firmly switched off when the export of capitalism has taken precedence over the spread of democracy.

Capitalism and liberal democracy have not been the only values that the United States' communication infrastructure has imparted to the rest of the world. Less deliberate, but even more effective, has been the transmission of American life-styles. If the incidence of democracy is a measure of the West's political impact and the pervasiveness of capitalism is a measure of its economic impact, then the spread of Western life-styles serves as a measure of its cultural impact. Indeed, there has even been a competition within the Western tradition itself for influence in the Third World, a rivalry between Europe's ancestral culture and America's cultural revisionism. Europe's cultural exports to the Third World are generally distinguished from those of the United States by their finer qualities. European style and sensibility have been conveyed through fine art, high fashion, classical music, university structures, prestige newspapers, and culinary traditions, among others.

In most cultural arenas, however, American genius lies in the popular form rather than the elite specialty, in mass involvement rather than aristocratic cultivation. Thus, America's jazz is better known than its classical music, its news magazines are better known than its novels, its casual dress more appreciated than its formal stylings, its fast food more admired than its formal cuisine, its soft drinks better known than its alcoholic beverages, its dramatic television series more appreciated than its public affairs programs, and so on. Alexis de Tocqueville would feel abundantly vindicated. America was the West's first mass democracy. Why should its popular culture not be its main claim to global immortality?

Unfortunately, America is at the same time largely oblivious to the popular culture of the rest of the world. Its collective genius for popular communication is not matched by its capacity for collective listening. For example, though American popular music has itself felt the impact of African and Caribbean rhythms, this influence has almost never come directly from those regions. Access to African music—or, for that matter, African literature, art, food, or clothing—is nearly impossible outside New York, Washington, and one or two other American cities with sizable Caribbean and African immigrant populations. On the other hand, obtaining American cultural exports in Kampala, Lagos, Nairobi, or Kinshasa is quite easy. The world has learned to dance to the music of the United States, but America has yet to listen to the concert of the world.

What lies behind this paradox of the American condition? Why is America an effective communicator but an inattentive listener? Why is the world becoming increasingly Americanized while America fails to be simultaneously humanized?

It is first necessary to clarify what precisely the humanization of America might entail. Four conditions would need to be fulfilled: the deracialization of America's world view; enhanced American tolerance of global cultural pluralism; greater American sensitivity to the economic needs of other nations; and the suspension of American political and military intervention in the Third World.

Race and culture have deep roots in America's history and outlook. Race and religion, in particular, have played key roles in shaping the character of American democracy. One could argue that American democracy was born out of religious toleration, on the one hand, and racial intolerance, on the other. Religious toleration provided a foundation for an ecumenical attitude toward the world's cultures, but racial prejudice served to neutralize America's nascent responsiveness to cultural pluralism.

While it is indeed true that the Pilgrims turned out to be greater zealots than the religious persecutors they had fled in Europe, the overarching trend in American history has been toward greater religious tolerance. By the time its constitution was being drawn up, the United States had already surpassed Europe in the effort to divorce the state from the church. The federal charter formally proscribed laws which infringed on the freedom of worship or enhanced the political status of one denomination over others. The secular state in Western history was at hand, and it held considerable promise for the accommodation of diverse cultures and creeds.

Yet the same America which was learning to be more religiously tolerant than Europe was, at the same time, learning to be less racially tolerant than its forebears. Paradoxically, America's relative religious enlightenment coexisted from the outset with a measure of relative racial intolerance. In a sense, Europe contributed to this paradox, for it was the trans-Atlantic slave trade with Europe's American colonies which helped to institutionalize racism in the Western Hemisphere.

By the time America was engaged in the process of creating a modern secular state, it had already begun to fail in the process of creating a nonracial society. The judicial principle of "separate but equal," established in a series of Supreme Court decisions after the Civil War, served to intensify the racial inequality of American society. There had been a contest between God and genes. American democracy assiduously sought to keep God out of politics, but it retained genetic discrimination within its political processes. The Founding Fathers had avoided establishing a theocracy, but neither they nor their nineteenth-century successors seemed as concerned about the dangers of ethnocracy. American democracy was less sacralized but more racialized than its European counterparts.

Abraham Lincoln dramatically embodied America's paradoxical fusion of spiritual vision and racial blinders, its religious virtue and ethnic vice. In that very contradiction Lincoln also manifested his nation's torment. Lincoln continues today to be widely regarded by Americans as an eminent symbol of Christian compassion and democratic sensibilities. Although it is true that his religious piety turned him against slavery, Lincoln's racial prejudices also turned him against anything approaching equality between whites and blacks. In a speech on September 18, 1858, in Charleston, Illinois, Lincoln made it abundantly clear that he was for emancipating "the Negro" but not for embracing him. He said yes to Negro freedom but no to Negro equality:

> I do not understand that because I do not want a negro woman for a slave I must necessarily want her for a wife [cheers and laughter]. . . . I will to the very last stand by the law of this state, which forbids the marrying of white people with negroes. . . . I will say then that I am not, nor ever have been in favor of bringing about in any way the social and political equality of the white and black races [applause]—that I am not nor ever have been in favor of making voters or jurors of negroes; nor of qualifying them to hold office, nor to intermarry with white people.[1]

This contradiction between high religious or moral purpose and racial prejudice, between democratic ideals and ethnic exclusivity, continues to be an American dilemma. Certain habits are ingrained. Thus, when Carter was president, he and his black ambassador to the United Nations, Andrew Young, both Baptists, still prayed in separate white and black churches in Georgia. Americans do not understand the impression this conveys to the wider world; that lack of understanding is one part of the U.S. insensibility to outsiders, part of the phenomenon of its deaf ear.

If the United States is at once religiously liberal and racially bigoted, how is this contradiction manifested in American foreign policy in the Third World? It turns out that it is not just America's respect for cultural diversity that has been distorted by race, but also its economic generosity and the use of its military power. The most generous things white Americans have done have been directed toward fellow white people; the most cruel, toward nonwhites. American generosity perhaps reached its apex with the Marshall Plan. American cruelty has ranged from genocide against native Americans to the dropping of atomic bombs on Hiroshima and Nagasaki, from the lynching of "niggers" in the southern states to the war in Vietnam.

Under the Marshall Plan, the United States spent nearly $20 billion on European reconstruction, an extraordinary sum at the time. It is difficult to imagine the U.S. Congress approving anything near this sum of money for the rescue of contemporary nonwhite societies. In the 1980s Africa needs a rescue operation on the scale of the Marshall Plan, yet the United States has in recent years consistently directed a much smaller proportion of its annual GNP to development assistance than have its advanced industrial allies.[2]

Because it is the home of many survivors of the Nazi Holocaust, Israel has also been the beneficiary of American largesse. In the postwar era, the United States has already spent more on Israel than on the entire Marshall program. Few outside the Arab world would begrudge this assistance, if it were not for the contradiction between America's support for Israel and its relative lack of interest in its own domestic minorities or in the wider world of nonwhites. While the Reagan administration has reduced federal outlays for welfare and other social programs within the United States, it has at the same time increased military and economic support for a Jewish community outside the United States. A foreign country with a population of less than four million has been getting disproportionately more of America's attention than the domestic black minority of thirty million people. The contrast raises questions that have been uncomfortable for the United States to confront.

The United States has perfected a special censorship to prevent itself from hearing distressing world voices. Some are born deaf, others become deaf, and some inflict deafness upon themselves.

The messages from abroad that the United States has been least prepared to listen to during the postwar era are those of Marxism and Islam. In the Third World, Marxist opposition to the United States mainly concerns the issue of economic imperialism, ranging from the omnipresence of American multinational corporations to the preeminence of the U.S. dollar in international economic relations. Islamic reservations about the United States are focused more on American cultural imperialism.

Economic imperialism is the sustained use of economic power or stratagem by a stronger state to control or penetrate the economy of a weaker country in the ultimate interests of the stronger. The financial advantages unfairly pursued in this manner have traditionally included access to markets, opportunities for private investment, availability of raw materials, and access to cheap labor. Cultural imperialism is the undermining or "polluting" of the values, mores, institutions, and sense of identity of another society, either by enforced assimilation into a proselytizing culture or as a side effect of other forms of domination such as economic imperialism or military occupation.

Third World Marxists do not want their economies to be controlled or exploited by the West; Third World Muslims seek to arrest the contamination of their cultures by the West. To Third World socialists, Marx is pitched against Uncle Sam in a struggle for economic self-determination. To Third World Muslims, Muhammad confronts Uncle Sam in a struggle for salvation and cultural purity.

From the perspective of the United States, the two struggles are often confused, but they are actually somewhat different. The struggle against Marxism is strategic, with religious overtones. The struggle against Islam is racial and religious. Marxism poses its challenge across an East-West divide, which shapes up primarily as a civil war among whites. Islamic "fundamentalism" confronts the United States across a North-South divide which is often manifested as a racial war of white against nonwhite. The United States perceives Marxism in the Third World as an extension of its confrontation with the Soviet Union. A certain quality of sanity and stability is attributed to East-West relations, while relations with Islamic peoples are assumed to be uniquely complicated by the Muslim world's "fanaticism" and "instability."

Islam is a predominantly Afro-Asian religion. To that extent, it is a religion of nonwhite people (which is precisely why it has sometimes fascinated black Americans from Malcolm X to the boxer Muhammad Ali). For white Americans, the crusade against the presumed threat of Islamic fundamentalism is partly a struggle against the forces of nonwhite self-assertion and autonomy.

The United States refuses to consider that Third World Marxism, far from being an extension of East-West tensions, is in fact a manifestation of North-South unease. People in the Southern Hemisphere are radicalized not because they are anti-Christian or even anticapitalist, but because they are primarily anti-imperialist. Scratch a Third World Marxist and you will find a Third World nationalist. The Ugandan radical analyst J. R. Barongo illustrated this reality when he wrote, "Many of the political problems existing in African countries today are predominantly external in origin . . . largely explicable in terms of material poverty and the dependent nature of African countries on the operations and manipulation of the international capitalist system."[3]

In the final analysis, then, Marxists' hostility is directed not toward capitalism as a form of economic and social organization, but toward imperialism as a means of external domination. In Africa, for example, the local bourgeoisie are certainly much less threatening to mass interests than are American corporations. The most significant political cleavages in the Third World are not really between national classes but between international power blocs. Third World Marxists are anti-American more because America is a world power controlling their economies than because America has a capitalist mode of production. Marxism in the Third World is a radical version of nationalism, though the United States does not recognize it as such.

America has been similarly undiscriminating in listening to messages from the Muslim world. Despite what many in the United States and elsewhere believe, the Iranian revolution was not anti-Christian, but anti-Western. The chief focus of hostility was not the Vatican, but Washington; not the crucifix, but the star-spangled banner. Iranian motivations were no doubt religious, but the targets were secular. The revolution ought not be seen as an ancient crusade involving contemporary equivalents of Saladin and Richard the Lion-Hearted; it was a modern crusade involving the muezzin from the minaret in a confrontation with disc jockeys, the Ayatollah in opposition to the White House.

Marxism and Islam have shared the common goal of trying to prevent the Americanization of the world, but they have parted company in the effort to humanize America. Marxism has attempted to prevent further American penetration of the world economy essentially by rallying the resistance of workers and peasants. Islam has sought to oppose American cultural encroachment by propounding the ideals of Third World authenticity and cultural dignity. However, for all their considerable success in slowing the Americanization of humanity, neither Marxism nor Islam has made much progress in fostering the humanization of America. An effective method of operating America's hearing aid has yet to be discovered.

Hearing aids can be switched on or off, put on or taken off. In the twentieth century the United States has been much more prepared to proclaim to the Third World than to respond, much more prepared to articulate than to listen.

As we have seen, this can in part be explained by the religiously liberal but racially prejudiced character of American society. A fundamental prerequisite for the resolution of America's estrangement from the Third World will be its progress toward humanization. A liberal political culture gave the United States a good start in respecting cultural pluralism, but the persistence of race consciousness in the United States prevented that tolerance from attaining fulfillment. Racism has also interfered with America's responsiveness to the economic needs of others and with its capacity to restrain the deployment of its own political and military power in the Third World.

It would not be unreasonable to expect, for example, that the United States would have strong ties with the west African nation of Liberia, for that country was founded in the 1830s by freed American slaves who modeled their republic after the U.S. system. It is therefore instructive to consider the vast disparity between America's economic and political commitments to Liberia and to Israel, in whose well-being the United States has always taken a special interest. What explains this paradox? Three factors which have often been advanced to explain the closeness of U.S.-Israeli ties—the integration of the two nations' economies, kindred political systems, and linguistic and cultural affinities—are no less pertinent to the Liberian case.

Furthermore, while it is true that American strategic interests are far more important in the Middle East than in west Africa, the United States has never provided comparable support to its various Arab clients. Its support for Israel has sometimes harmed rather than protected America's vital interests in the region, and these interests are even less vital than those of its NATO allies, which have developed more flexible Middle East policies. It is at least conceivable, then, that white Americans generally feel closer to the Israelis than they do to the Liberians because, when all is said and done, Israel is European-led while Liberia is a black nation. Might it not also matter that Israel's main advocates in American public and private life are overwhelmingly white, while influential supporters of black Africa, when they can be found, are predominantly black?

In light of these considerations, America's closeness to Israel has greater relevance to U.S. political relations with the Third World than is often realized. Clearly the U.S.-Israeli relationship affects U.S. ties with the Arab nations, an increasingly influential part of the Third World. It also affects America's relations with the rest of the Muslim world. Israel's connections with South Africa have, in turn, alienated many Africans, while its arms sales to right-wing movements and regimes in Latin America—Israel has reportedly been ready to step into the breach when the U.S. Congress has prevented the executive branch from providing military support—have cost it support among Latin liberals and radicals, both lay and clergy. The effects of Israel's relative isolation from the Third World mainstream have been more than compensated for by its friendship with the United States. But that very friendship, remarkable though it may be, must also be recognized for its contribution to the joint estrangement of Israel and the United States internationally.

9 The Adams Doctrine and the Dream of Disengagement

PHILIP L. GEYELIN

Philip L. Geyelin, a syndicated columnist in Washington, is editor-in-residence at the Foreign Policy Institute of the Johns Hopkins School of Advanced International Studies. He is a former diplomatic correspondent of the *Wall Street Journal*, and from 1968 to 1979 he was editorial page editor of the *Washington Post*, a role in which he was awarded the Pulitzer Prize. Mr. Geyelin is the author of *Lyndon B. Johnson and the World*.

> Unless you draw back from your present course of action, this will inevitably jeopardize the course of United States–Soviet relations throughout the world. I urge you to take prompt constructive action to withdraw your forces and cease interference in Afghanistan's internal affairs. —President Jimmy Carter, by "hot line" to Leonid Brezhnev, immediately following the Soviet invasion of Afghanistan, December 27, 1979

IN HIS MEMOIRS, Jimmy Carter was to call this the "sharpest" message he ever sent the Soviet leader during his presidency. In the memoirs of Cyrus Vance the chapter on Afghanistan is subtitled "A Turning Point in U.S.-Soviet Relations." Vance thought the administration's response—a grain embargo, a ban on the sale of high-technology equipment, restrictions on Soviet fishing privileges in American waters, postponement of the opening of new consulates in Kiev and New York, and withdrawal from the 1980 Olympics in Moscow—was "strong and calculated to make Moscow pay a price for its brutal invasion." In his State of the Union message of January 1980, Carter proclaimed what was to become the "Carter Doctrine": "An attempt by any outside force to gain control of the Persian Gulf region will be regarded as an assault on the vital interests of the United States of America, and such an assault will be repelled by any means necessary, including military force." Some thought Carter had to be bluffing, given the difficulty of moving American forces into position rapidly enough to stop a Soviet military thrust in the region. But the United States did strengthen its forces in the Persian Gulf. And Carter insists in his memoirs that the U.S. response to a Soviet assault in the gulf would "not necessarily [have been] confined to any small invaded area or to tactics or terrain of the Soviets' choosing."

Neither the actions nor the rhetoric was enough for Ronald Reagan, then a presidential candidate. He advocated halting all trade with the Soviet Union and urged the "Western World" to join this "quarantine" until the Soviets "decide to behave like a civilized nation." He proposed a moratorium on all treaties with the Soviet Union, as well. At another point he spoke of blockading Cuba to "stop the transportation back and forth of Russian arms, of Soviet military." When George Bush called such talk "a lot of macho," Reagan only became more precise: "Suppose we put a blockade around that island and said, 'Now, buster, we'll lift it when you take your forces out of Afghanistan?' "

In the aftermath of his crushing 1980 election victory, Reagan's hard-core conservative constituency contrived a "mandate" out of this and similar Soviet-bashing, a mandate for a massive defense buildup in support of a bare-chested, rough, tough policy to turn back the communist tide worldwide. And yet, George Bush was right: It was "a lot of macho." By election day 1984, although the U.S. defense buildup was on track, the Soviets were still in Afghanistan. There had been no quarantines or blockades. American grain was flowing freely to the Soviet Union, since Reagan had lifted the embargo in fulfillment of a campaign pledge to the farmers. Afghanistan still topped the Republican foreign policy bill of particulars, but there is no evidence that it was anywhere near the top of the list of electoral concerns—not as a continuing demonstration of Soviet brutality, not as a security threat to the United States, and not as an indictment of the "Carter-Mondale administration."

Afghanistan, by reason of its physical inaccessibility to news reporters, was out of sight and largely out of mind; that may say something about the role of the media in establishing the American foreign policy agenda. But the comparative record on Afghanistan of the Carter and Reagan administrations provides a perfect point of departure for the examination of various elements at home and abroad which confound the conduct of a coherent, consistent American foreign policy.

You can blame it, if you like, on the emergence after World War II of an increasingly complex, essentially unmanageable world, transformed by the appearance of a multitude of new states emerging from colonial rule. Most were unprepared to govern in a stable way. Many lacked the wherewithal of natural resources and were forced to organize societies within arbitrary, often contested borders. Or you can blame it on technology: The Bomb made it unthinkable to try to settle even the bitterest quarrels between the great powers by resort to the ultimate means. From this derived the need for a new concept of "limited" warfare by limited means which, when carried to their logical extreme, find terrible expression in terrorism. You can find "the focus of evil in the modern world" in the Soviet Union. Or you can opt for a more complicated view that takes into account everything from Islamic fundamentalism to ancient tribal or secular animosities, grinding poverty, social imbalances and consequent injustices, and population growth that outpaces food production.

Better yet, you can accept all of the above and conclude that an American yearning for separation from such a complex, frustrating world would be understandable. But if you are among the many who think there is no escaping the world as it is, and that to try to do so would be unacceptably dangerous, it is not enough to look outward for an explanation of this yearning to escape. It becomes necessary to look inward as well for the roots of the estrangement that has emerged in public policies and public opinion, even as many politicians hail a patriotic revival and an America "standing tall" as a respected and influential force in the world.

Looking inward, it becomes clear that what now strikes many people as a reversion to semi-isolation from a traditional, intimate, and sustained involvement in the world's affairs can more properly be seen the other way around. The United States is now experiencing a return from a particular period of intense and untraditional involvement beyond its shores in the two decades after the end of World War II to what is America's natural and historic condition: detached and wary of entanglement, content to nourish cultural ties and ancestral connections, pleased to profit from international commerce, but preoccupied with perfecting the American way of life, liberty, and the pursuit of its own economic well-being. Detachment—or, at the most, rare moments of engagement—is America's natural state.

This is not to deny that at various stages American leaders have seen urgent and valid reasons to intervene actively in faraway parts of the world, when American interests were thought to be threatened. This was the case after World War II, when the old European empires were collapsing and the United States had unquestionable military supremacy, a nuclear monopoly, and was economically overpowering as well. But more recent history shows that experience with an increasingly less manageable world, and with American power diminished in relation to Europe as well as the Soviet Union, has tended to temper the world-policeman instinct of even the most adventuresome, conceptualist presidents and/or secretaries of state. Not the least of the reasons for this—leaving aside the prudence imposed by a nuclear balance of terror—is that the American democratic system does not lend itself to the sort of consistency that would give staying power to such a policy.

But a more fundamental factor is that the American public is not by nature inclined to world policing, and still less to imperialist actions. Even the most notable exception—Central America and the Caribbean region—reinforces the rule. From the Monroe Doctrine forward, successive American governments have adopted a proprietary interest in what is disparagingly described as America's "backyard." Depending on how you count assorted interventions of one kind or another, the United States since 1850 has involved itself in the internal affairs of the area immediately to the south some thirty times, with gunboats or marines or larger forces including the U.S. Army. But the motive was actually anti-imperialistic, whether Washington was concerned with the actions of Britain, France, Spain, or other Western imperialists of the late nineteenth and early twentieth century or, more recently, with the Soviets and the expansion of international communism. Just as the Reagan administration today would like to overthrow the Sandinista government in Nicaragua, so have past administrations acted reflexively to restore order, rescue Americans, protect friendly regimes, forestall the threat of uncongenial regimes, or overthrow hostile governments in Guatemala, the Dominican Republic, Nicaragua, Haiti, Honduras, Cuba, and Panama.

"The interventions of the United States in the Caribbean and in Central America," Samuel Flagg Bemis argues in his comprehensive history *The Latin American Policy of the United States,*

> were not impeccably carried out, but the dominating motive of those interventions and of the "dollar diplomacy" that is usually associated with the interventions was not the exercise of dominion over alien peoples—the hallmark of imperialism—nor their exploitation by an "economic imperialism"; it was to foster their political and economic stability so that there could be no justification, or pretext, for Euro-

> pean intervention in such a vitally strategic area of the New World. It was also, in the case of Haiti, to protect foreign citizens, including those of the United States, during a complete collapse of law and order there.

While others might give greater weight to the influence of large American commercial interests, and define it as undeniable "exploitation," most authorities would agree with Bemis that the American imperialist fling between, say, 1898 and the 1920s was

> comparatively mild. . . . A careful and conscientious appraisal of United States imperialism shows, I am convinced, that it was never deep-rooted in the character of the people, that it was essentially a protective imperialism, designed to protect, first the security of the Continental Republic, next the security of the entire New World, against intervention by the imperialistic powers of the Old World. It was, if you will, an imperialism against imperialism. It did not last long and it was not really bad.[1]

It took place, moreover, in a relatively modest "sphere of influence." If you set that sphere aside, there is not much else in the more than two hundred years of the republic to suggest more than an occasional, spasmodic interest in throwing American weight around. One may think of this phenomenon in terms of a shock cord, which can be stretched far beyond its natural, relaxed condition for specific purposes and under particular circumstances, but which, when released, snaps quickly back to its natural shape. The natural shape of American foreign policy was defined by George Washington in his Farewell Address in 1796:

> The Great Rule of conduct for us in regard to foreign nations is, in extending our commercial relations to have with them as little *political* connection as possible. . . . Why quit our own to stand upon foreign ground? . . . Taking care always to keep ourselves by suitable establishments on a respectable defensive posture, we may safely trust to temporary alliances for extraordinary emergencies.

With a little editing, this could be an excerpt from the debate on the War Powers Act of 1973 or on its application, ten years later, to the deployment of U.S. marines as part of a multinational force in Lebanon. In 1821 John Quincy Adams expanded upon this inheritance by setting forth a few first principles that—with some worthy exceptions, and some not so worthy—give to the negative notion of disengagement a rather more positive, not to say historically consistent, spin. The same words spoken today would be quickly converted for public consumption into an "Adams Doctrine" and be welcomed warmly in mainstream America:

> Wherever the standard of freedom and Independence has been or shall be unfurled, there will her [America's] heart, her benedictions and her prayers be. But she goes not abroad, in search of monsters to destroy. She is the well-wisher to the freedom and independence of all. She is the champion and vindicator only of her own. She will commend the general cause by the countenance of her voice, and the benignant sympathy of her example. She well knows that by once enlisting under other banners than her own, were they even the banners of foreign independence, she would involve herself beyond the power of extrication, in all the wars of interest and intrigue, of individual avarice, envy and ambition, which assume the colórs and usurp the standard of freedom. The fundamental maxims of her policy would insensibly change from *liberty* to *force*. . . . She might become the dictatress of the world. She would be no longer the ruler of her own spirit.[2]

Obviously, Adams could not have foreseen a world held in thrall by an overarching ideological conflict between two superpowers armed with enough weapons to destroy the planet, preaching and practicing altogether incompatible economic and social theories. He could not have envisaged fifty American states with a political center of gravity shifting ever southwestward, let alone a world with nearly 170 sovereign nations of myriad shapes, sizes, and political or secular persuasions. He could not have imagined the industrial and technological revolutions, nuclear fission or fusion, supersonic speed. He could not have begun to comprehend the political implications of modern communications and a volatile public opinion carried crazily this way or that way by symbols, shorthand, and imagery. Yet, in uttering these words, Adams had to know something about the enduring soul of the republic he helped to create and the spirit in which the United States was born, a spirit that would profoundly inform and define the American world role, even as American leaders themselves would defy, distort, or deny it for purposes that seemed right or even popular at a particular time.

As with George Washington, so it is with Adams: Loud echoes can be heard in recent debate. What, after all, was the intellectual argument against U.S. involvement in Vietnam if it did not have to do with the danger of going abroad in search of "monsters to destroy," if it did not have to do with the question of whether enlisting under the "banners of foreign independence" risked involvement "beyond the power of extrication." And what was the moral protest all about if not the specter of an America that "would be no longer the ruler of her own spirit"? With nineteenth-century elegance, Adams articulated what would later come to be known as the Vietnam syndrome. He was making the case against the Central Intelligence Agency's secret war in Nicaragua, not to mention the U.S. presence in Lebanon in the 1980s. He was saying something, as well, about the American public's response to Afghanistan: the initial outrage, the frustration, the strong support for the Carter reprisals—but a short attention span. The outrage gave way in due course to a readiness to return to doing business, as best one could, with the Soviets.

Adams was far too intelligent to lay down a rule without exceptions. World War II, for example, and also quite probably World War I, might have represented situations, even for Adams, in which America, the "well-wisher to freedom and independence," could hardly have ignored the threat to "her own." It is idle to wonder whether Adams or George Washington would have resisted the inflamed plunge into the Spanish-American War, the occupation of the Philippines, various Central American involvements, the intervention in Lebanon in 1958, the "liberation" of Grenada in 1983, or the long and costly Korean War. It is also beside the point. The Founding Fathers had an extraordinarily prescient, fingertip feel for the national character and condition. Americans would now and again be persuaded of a genuine threat requiring a counterthreat or, in some circumstances, the actual use of force. But with the exception of a few relatively cost-free, pulse-racing imperialistic flings, the rule has held up remarkably well. It has taken guile, a measure of demagoguery, and, on occasion, outright duplicity to mobilize Americans to do more than pay proud tribute to the superiority of their system and commend it, with the inducements of not entirely selfless economic and military aid, to others.

Even the extraordinary efforts made in the immediate postwar period to contain the Soviet Union, reconstruct Europe, and help the underdeveloped world were exercises sold to the American public on the grounds that they would introduce an order and stability that would make active U.S. engagement as a peacekeeper or as an intervenor less, rather than more, likely. In trying to make the world a safer and more prosperous place, America was declaring that, apart from flourishing commercial and cultural contacts, it would like to put itself in a position not to have to worry about everyone else; it was simply not the natural impulse of the American people to do so.

What explains this restraint? The analysts whose business it is to examine the collective American temperament and study polling data have ready answers. Americans are a mixed lot, ethnically and geographically, afflicted by more than the usual inner contradictions: national pride balanced by a natural prudence; a sense of physical remoteness provided by two great oceans, but also ancestral connections that reach across them; an East Coast oriented toward Europe and a West Coast looking out upon a Pacific Basin, with a landlocked heartland in between that seems far from either shore.

A tradition of leadership by example, as distinct from force of arms, runs back to the very beginnings of the nation. In 1630 John Winthrop, governor of the Massachusetts Bay Company, delivered what might be called the keynote address to the passengers of the *Arrabella* as they approached the New World. "For we must consider that we shall be a city upon a hill," he said. "The eyes of all people are upon us."[3] (Ronald Reagan's political oratory would make it a "shining city" three and a half centuries later.) America was to be, then, not so much the world's leader as the world's role model, and the durability of this prescription is regularly confirmed in the current estimates of political pulse takers that roughly two-thirds of the American electorate normally takes no interest in foreign policy. That is the consistent finding of election exit polling and public-opinion sampling.

William Schneider, a recognized authority on American public opinion and its relation to politics and foreign policy, thinks most Americans want the United States to be the "toughest kid on the block," but in a "defensive" way, as distinct from wanting to be a "world policeman." The perfect state is "peace and strength," he says. Which matters the most? "That's like asking which worries you the most, your health or your financial security," he replies. "The answer is both, if you're in the hospital. But the emphasis ebbs and flows with events."

The Schneider "ebb and flow" theory can be vividly demonstrated in quadrennial surveys by the Chicago Council on Foreign Relations entitled "American Public Opinion and U.S. Foreign Policy." According to data gathered in 1974, the American public was beginning to worry increasingly about a "perceived growing military imbalance between the United States and the Soviet

Union." This became a "major obsession following the events in Iran and Afghanistan at the end of 1979 and it played no small role in the 1980 presidential election," the council later found. Yet, somewhere along the way between 1978 and 1982, the public reversed its priorities. In 1978 more people wanted to expand defense spending than to cut it back. By 1982, seemingly secure in the knowledge that Ronald Reagan had made good on his 1980 campaign promise to rebuild American defenses, more people wanted to cut back military spending than to increase it. The 1982 survey showed increasing public concern about the state of the economy, budget deficits, and the threat of nuclear war.

The Chicago Council's analysis of 1982 polling results explicitly showed the Adams Doctrine at work in "a continuing erosion of the post–World War II public consensus that the national interest requires active participation by the United States in world affairs." In a report issued in 1983, it said, "Only a bare majority of the public now holds the opinion that such international activism is best for the future of the country while over a third now say it would be better if the United States 'stayed out' of world affairs."

This inherent reluctance to get involved can be measured in another way: the constituency for foreign aid. There is a persuasive argument that development aid is the ounce of prevention that is worth far more than the pound of cure that becomes necessary when the grinding poverty and economic and social injustice in the Third World lead to mass unrest or insurrection. With revolutionaries more susceptible by the nature of their grievances to co-option by the Soviets and Marxism-Leninism than by Western capitalism, the United States has invariably arrived with too little, too late—and, more often than not, on the losing side. The spirit that led the United States to reconstruct a war-shattered Europe out of its own enlightened self-interest, under the Marshall Plan, has not been matched by a comparable level of self-interested support for the underdeveloped world; the seedbeds of upheaval and potential East-West conflict have not been tended. The United States now ranks sixteenth among the industrialized nations in the resources as a percent of GNP that it allocates to economic aid.[4] Military aid has sometimes been more popular but, even so, has also been too little, too late, and often on the wrong side.

"Why are we so fatigued?" John F. Kennedy once exclaimed about the lack of American public interest in the crisis that followed the independence of the Belgian Congo in 1960. But it wasn't fatigue, really; it was the Adams Doctrine at work, a piece of a pattern in which the Afghanistan experience is only one of the most recent pieces.

If one examines the history of American foreign policy in the twentieth century, the Afghanistan crisis and the American response to it fall into place, underlining a certain consistency in the American spirit. A case can be made that Theodore Roosevelt was responsible for the most conspicuous departure from the Adams Doctrine over any prolonged period of time. He is remembered for the Panama Canal, San Juan Hill, and the slogan "Speak softly and carry a big stick." Teddy Roosevelt did not, in fact, speak all that softly, having once said, "No triumph of peace is quite so great as the supreme triumphs of war." But he certainly did, for a time, carry a big stick. He rose to power in the intoxicating times of American "manifest destiny." His predecessor, William McKinley, had seized upon the blowing up of the battleship *Maine* in Havana Harbor in 1898 to lead the United States into the Spanish-American War; Roosevelt, an assistant secretary of the Navy at the time, did some of the loudest drumbeating.

When the question arose of what to do with that war's spoils, even the mild McKinley was arguing in 1899 that, in the case of the Philippine Islands, "there was nothing left for us to do but to take them all, and to educate the Filipinos, and to uplift and civilize and Christianize them, and by God's grace do the very best we could by them as our fellow men for whom Christ also died."[5] Senator Albert J. Beveridge of Indiana was somewhat more explicit in a speech on January 9, 1900, on the Senate floor. "Self government and internal development have been the dominant notes of our first century," he said, adding:

> Administration and the development of other lands will be the dominant notes of our second century. . . . [God] has made us [our race] the master organizers of the world to establish system where chaos reigns. . . . He has made us adepts (*sic*) in government that we may administer government among savage and senile peoples. . . . And of all our race, He has marked the American people as His chosen Nation to finally lead in the regeneration of the world. This is the divine mission of America, and it holds for us all the profit, all the glory, all the happiness possible to man.[6]

The tone of that speech, Mark Sullivan wrote in the first volume of *Our Times*, a five-volume popular history of early-twentieth-century America, was "in the spirit of the times." He described it as "a key-note of Republican administration policy and platform." Later, in that same spirit, Roosevelt, as president, would first arrange for Panama's independence from Colombia and then engineer the construction of a canal through the Central American isthmus.

But there it ended. Although Teddy Roosevelt talked a good imperialistic game, projecting the promise of America as the arbiter of a stable world, his annexation of the Canal Zone actually marked the beginning of the end of American adventurism as a public policy—in the sense of an acquisitive, imperialist power reaching out for physical control over territory beyond its shores. The Adams Doctrine once again came to the fore. "The entire history of American overseas expansion is compressed practically within the year 1900 and the two years preceding," Sullivan writes in *Our Times*:

> We took the path of Senator Beveridge's "imperial destiny" far enough to complete the commitments we had become involved in, as unanticipated and rather disturbing incidents of a war we had begun only to rescue the Cuban people from the cruelties of Spanish rule. Then we stopped. For the change of emotion we went through, the cooling down from a hectic and rather artificially stimulated ardor, there is no phrase that conveys the picture quite so precisely as the slang one: we concluded to forget it. The American people, after about 1902, not only had a distinct disinclination for further expansion but were inclined to regard the annexations we had already made as embarrassing liabilities. By 1919 so great was our disinclination for responsibilities that when they were strongly urged upon us by other nations in the form of "mandates" over some of the former German dependencies, we rejected them with an emphasis that was clear evidence of an overwhelming and definitely crystallized public opinion in favor of America remaining at home.[7]

Americans instead turned inward, to the invention of the airplane and the Model T, to the first oil strikes, and to the emerging issues of civil rights, the organization of trade unions, and antitrust.

Out of World War I came the concept of the League of Nations, an ill-fated innovation that was entirely consistent with the American dream of a world that would not require too much attention. The same dream drove Franklin D. Roosevelt, a generation later, to embrace the United Nations, with all its promise of world order, even as World War II raged on. If the Truman Doctrine sounded like a prescription for a world policeman's role, the sweep of its commitment, Dean Acheson tells us in his memoirs, owed much to the political imperative of winning the support of a small but fiercely anti-communist minority in Congress for an aid program to Greece and Turkey.

Even so, the ideological evolution of that time baffled journalist Walter Lippmann when he sought some explanation in 1965 for the growing U.S. entanglement in Vietnam. "The problem of our foreign policy today," he wrote in a column,

> will not be fully understood until historians explain how our intervention in the second world war to defeat the Nazis and the Japanese became inflated into the so-called Truman Doctrine of the late 1940s, in which the United States said it was committing itself to a global ideological struggle against revolutionary communism.
>
> For it is this global commitment which is at the root of our difficulty in appraising coolly the extent and the importance of our engagement in Viet Nam.

Lippmann was not preaching pure Adamsism. But he did think there had to be a "stopping point between globalism and isolationism." As examples of cases where ideology had given way to simple prudence, he cited the restraint the United States had shown on the occasion of the Berlin uprisings in 1953, the Hungarian insurrection in 1956, and the Chinese conquest of Tibet in 1965 ("as naked a case of aggression as any we have seen"). In the matter of Tibet, Lippmann had an explanation that would have served as well fourteen years later when the Soviets invaded Afghanistan. The United States did not intervene in Tibet, he wrote, "for the good and sufficient reasons that the United States could not reach Tibet in order to defend it, and what is more, that Tibet is not a vital interest of the United States for which the President has the right to spend American lives." Tibet, then, was a "stopping point," and the test of statesmanship, Lippmann concluded, is to "find those stopping points and to act accordingly."[8]

Harry Truman did not see such a "stopping point" when he took the U.S. into the Korean War. But he was fearful enough that others might see one. In part to avoid the risk of a paralyzing domestic debate, he chose not to seek a formal declaration of war. Later, Dwight D. Eisenhower, correctly reading the public mood, swept to a landslide in 1952 with a promise to "go to Korea"—to go there, that is, to end the war.

John F. Kennedy is best remembered by many for the ringing passage in his Inaugural Address that begins, "We shall pay any price, bear any burden. . . ." It is all but forgotten that the same address also offered the more reserved vision of a "long twilight struggle" against assorted evils threatening humankind, in which the Soviet Union was invited to participate as a partner. After the Bay of Pigs, the Cuban missile crisis, and two and a half years of on-the-job training, Kennedy delivered a speech at American University in 1963 that defined as a proper American objective not the Wilsonian world "made safe for democracy" but, more modestly, a world "made safe for diversity."

Two months before his assassination, Kennedy was asked by Chet Huntley and David Brinkley of NBC why, with all of America's prestige and money, Washington could not exert more influence in South Vietnam. His answer was more in the spirit of Adams than of the turn-of-the-century American imperialistic fling:

> We have some influence, and we are attempting to carry it out. I think we . . . can't expect these countries to do everything the way we want to do them. They have their own interest, their own personalities, their own tradition. We can't make everyone in our image, and there are a good many people who don't want to go in our image. In addition, we have ancient struggles between countries. . . . We can't make the world over, but we can influence the world.

What he was concerned about, Kennedy said, was that Americans "will get impatient and say because they don't like events in Southeast Asia or they don't like the government in Saigon, that we should withdraw. That only makes it easy for the Communists. I think we should stay."[9] And so America stayed. But Kennedy was right in his fears about American impatience, as Lyndon B. Johnson was to discover in the hardest way.

Johnson had a master politician's sense of American gut feelings. He rode to a landslide in 1964 by playing on the alleged trigger-happiness of Barry Goldwater and by soothing public anxieties with assurances that American boys would not be sent to fight wars that Asian boys ought to fight for themselves. By the time he was through with his escalation of the war (and with his own political career) in March 1968, Johnson had built the American combat forces in Indochina up to more than half a million; his chokepoint was the additional two hundred thousand that the Pentagon was seeking at the time.

Vietnam had many lessons, and not the least of them was that Johnson was too clever by half in conning Congress and the public. The Tonkin Gulf Resolution, the trumped-up pretext that gave him sweeping authority to wage war; his careful concealment of the real implications of the first landing of American combat units in 1965; his refusal to test the public's deepening concern by instituting an equitable draft-deferment policy or calling up the reserves; his insistence on not calling the war a war or making the Congress a willing and responsible partner to it—in all these ways Johnson violated the Adams Doctrine, though he was fully aware of the strain in the American character that impelled Adams to propound it in the first place.

In the 1968 campaign the Vietnam War was Vice-President Hubert Humphrey's albatross just as Richard Nixon's promise to end the war was for many Americans the most appealing thing about him. Nixon was attacked for having a "secret plan" to "win" the war, but all he actually said he would do was to "end the war and win the peace in the Pacific." Whatever that meant, it was enough to see him through the campaign on the Vietnam issue. But Nixon's pledge lost its magic once he tried to put his plan into practice. When it developed that he had no better solution than Lyndon Johnson, even though he began to withdraw American combat troops, the antiwar protesters turned on him as they had on Johnson, with the same fervor and with the same effect on the war effort. An open society, freedom of expression, the right to dissent—all the values that Adams had warned would be put at risk if the United States sought to be the "dictatress" of the world—worked against a limited war. Broad domestic support for American policy would have been required to deter the leadership in Hanoi from continuing a decades-old campaign to unify their country by force. But there was no broad support for U.S. involvement in Vietnam and, therefore, no signal of American staying power was ever sent.

The values Adams prized were put at risk, as Richard Nixon embarked on his desperate effort to stifle dissent within the government and among the public, unleashing lawless "plumbers," burglars, wiretappers, and compilers of "enemies lists." Watergate was the proximate cause of Nixon's resignation in disgrace, as certain impeachment by the House loomed, but Watergate was the offspring of Vietnam and Nixon's frustration with that war.

On a loftier level, even while wrestling with Vietnam, Nixon recognized how far the Vietnam debacle had carried American foreign policy away from mainstream, traditional American policy. American defense strategy was scaled down from a capability to fight two and a half wars all at once to a more modest contingency plan for one and a half.[10] With the geopolitical conceptualist Henry Kissinger at his elbow, Nixon propounded a doctrine designed to preclude other "Vietnams" by placing greater responsibility on Asians for their own defense, while circumscribing the conditions under which the United States would intervene.

Even the Paris Accords and the withdrawal of the last American combat forces in 1973, however, did not suppress the emerging Vietnam syndrome. That same year a once-bamboozled and still embittered Congress passed the War Powers Act, putting strict constraints on the circumstances under which a president could commit U.S. forces in "hostile environments" without congressional consent. The act was passed over Nixon's veto and is still considered by its critics to have been an aberrational and perhaps unconstitutional congressional invasion of presidential power and prerogatives. The Vietnam syndrome itself is still regarded by many as an aberrational overreaction to the trauma of Vietnam. But subsequent events would reinforce the view that Vietnam was the exception and the syndrome an expression of a natural and historic inclination, which achieved special force on this occasion only because of the shock and anguish associated with Vietnam.

Gerald Ford was not in office long enough to prove much of anything, although two incidents during his administration help define the collective inner conflict that has characterized the American public's approach to the world through much of the last two centuries, a conflict that transcends the transitory trauma of Vietnam. The "toughest kid on the block" variation from the norm was illustrated by the surge of national pride that followed the bungled and costly rescue of the American crew of the cargo ship *Mayaguez* from Cambodian piracy only a month after Saigon fell to the Communists in April 1975. (The forty-man crew was recovered unharmed, but during the assault on their Cambodian captors eighteen Americans were killed or missing and presumed dead and fifty more were wounded. In addition, another twenty-three U.S. servicemen were killed when the helicopter transporting them to the area of the rescue went down in Thailand.) That the American cheering was out of proportion to the accomplishment can be credited to the post-Vietnam collapse of American self-confidence. But it has to be credited as well to something deeper and more enduring: If you accept the "toughest kid" theory of the case, the toughest kid doesn't necessarily prove his mettle by looking for trouble; he proves it by not allowing himself to be pushed around.

Thus Americans can throw their hats in the air over something like the recovery of the *Mayaguez,* or even something as trifling as the shooting down of two Libyan jets over the Gulf of Sidra. The flip side of the same theory can be found in the extraordinary congressional reaction when Kissinger, then Ford's secretary of state, proposed a covert CIA operation in Angola in 1975. The passage of the Clark Amendment denying funds for that purpose has usually been credited to the Vietnam syndrome; but it, too, reflected a deeper, longer-standing aversion to go out looking for trouble in remote and unfamiliar corners of the world, where the toughest kid's toughness is not being put to any dramatic or demonstrable test.

The fates might have been kinder to Jimmy Carter if the chronology of his term as president could somehow have been reshuffled, putting the last year first. He might then have gone into a campaign for reelection with the hostages long since recovered and the Soviet brutality in Afghanistan a more distant memory. He would have been the "benignant" peacemaker of Camp David, in the Adams tradition, and the champion of a strong defense, agreeing with the Europeans to deploy intermediate-range nuclear missiles in the absence of an arms limitation agreement with the Soviets. He would have been the bold political risk taker who had negotiated and fought for a modern, realistic Panama Canal treaty. He would have been credited with making peace with the world's most populous country by normalizing relations with its communist government, a far more difficult accomplishment, politically and diplomatically, than Nixon's breakthrough visit. That is a whimsical hypothesis, but it makes a point: Events beyond the mastery of any American president can interact powerfully with the relentless and immutable U.S. electoral clock by abruptly and profoundly altering the appearance of things, and thus altering the outcome of elections in a way that distorts the underlying and enduring American view of the world and the U.S. role in it.

Carter shrewdly played to first principles in his 1976 campaign and in the first years of his administration—indeed, in his first major foreign policy speech, at the University of Notre Dame in 1977. There he dwelt heavily on the role of American values in the pursuit of human rights abroad. Innocently, he left himself wide open to misrepresentation when he attempted to sum up public sentiment at the time: "Being confident of our own future, we are now free of that inordinate fear of communism which once led us to embrace any dictator who joined us in that fear." "Inordinate" implies excess, and excessive fear is a poor guide to policy. Carter was tapping into mainstream American thinking, harking back to John Quincy Adams, or so he believed. He acknowledged that American policy had been guided since 1945 by "a belief that Soviet expansion was almost inevitable but that it must be contained," and he didn't quarrel with that. He was simply saying that "for too many years we've been willing to adopt the flawed and erroneous principles and tactics of our adversaries, sometimes abandoning our own values for theirs. . . . This approach failed, with Vietnam the best example of its intellectual and moral poverty." He went on to say: "But through failure we have now found our way back to our own principles and values, and we have regained our lost confidence."[11] Years later, in the 1984 campaign, however, Republicans, exploiting a very different public mood, would get at Walter Mondale by citing such statements by Carter as evidence that the Democrats were weak and naive about the communist threat—soft, in other words. The implication of this familiar and effective Republican charge, heard in 1980 and again in 1984, was that these Democratic qualities had generated a new wave of Soviet expansionism during the Carter presidency.

It was not so much that Afghanistan, for example, was a black mark against Carter, in and of itself. It was simply one part of an apparent pattern that lent itself to political exploitation. Carter was ridiculed by the Republicans when he said on television that "this action of the Soviets [in Afghanistan] has made a more dramatic change in my own opinion of what the Soviets' ultimate goals are than anything they've done in the previous time I've been in office."[12] Conservatives said this was proof of Carter's naiveté, since the Soviets' goals, they believed, had never varied. But the president's remarks probably captured the reaction of most Americans who had not previously paid much attention to Afghanistan.

Still, there it was, just in time for the 1980 campaign: the image of a president so consumed by detail and so given to pieties that he couldn't get anything right. In contrast, there was the image of Ronald Reagan—unsupported by any record but reinforced by red-hot rhetoric—the all-American, star-spangled, red-white-and-blue embodiment of Uncle Sam, standing tall. As he later demonstrated, Reagan probably would have done nothing more decisive about Afghanistan, perhaps not even as much as Carter did, had he been in the White House at the time. But that did not deny Reagan the challenger's right to argue, to good effect, that Afghanistan, the Iran hostage crisis, and several other unfortunate developments would not have occurred if he had been in charge and the United States had not tempted Soviet adventurism by letting down its guard. Thus, Afghanistan was far more important for its immediate psychological and political impact in the 1980 election than for its long-term effect on U.S.-Soviet relations, or as a marker of some fundamental change in the collective American sense of the nation's proper role in the world.

The Soviet invasion of Afghanistan, in short, was not so much a turning point as an exclamation point. The exclamation was that a disorderly and ineffectual America had become incapable of containing communism, of creating the conditions conducive to East-West coexistence, of coping with an unruly world. That was the perception carefully cultivated by the Carter administration's most vociferous opponents: The rape of Afghanistan was of a piece with Soviet opportunism in Ethiopia, Angola, Yemen, Central America, and the Caribbean. Afghanistan spelled the end of détente.

What was the reality? Even within the Carter administration, with foreign policy debate polarized between the State Department and the White House (between Cyrus Vance and Zbigniew Brzezinski), there were different understandings of the facts, and a disagreement over remedies. Moreover, the chain of events that some believe link Afghanistan to Jimmy Carter's foreign policy actually stretches back into the Republican years of Richard Nixon and Gerald Ford. According to his memoirs, Kissinger was already saying eulogies for détente with the onset of the Watergate constitutional crisis in 1973. Kissinger's concept of détente required "linkage" of Soviet and American behavior across the board. It was "built on the twin pillars of resistance to Soviet expansionism and a willingness to negotiate on concrete issues," Kissinger wrote, asserting that "the discipline and sense of proportion necessary for such a course fell prey to the passions of the Watergate era." In a sense, Kissinger argues, détente killed itself: Conservatives were uneasy with its apparent "softness," while liberals turned against it out of their deep hatred for Nixon.

A particular passage of Kissinger's memoirs is worth citing at some length for its relevance to the Adams Doctrine. Noting that in the postwar nuclear world "statesmen now no longer risk their armies but their societies and all of mankind," Kissinger observed that adversaries had thus become partners in "the avoidance of nuclear war—a moral, political, and strategic imperative." He continued:

> No society has ever faced such a manifold task; few could have been less prepared for it. The only kind of threat to the equilibrium for which history and experience had prepared us—an all out military assault on the Hitlerian model—was the least likely contingency in an age of proxy conflicts, guerrilla subversion, political and ideological warfare. The modern challenges were ambiguous in terms of our expectations, were resisted hesitantly if at all, and—from Korea through Vietnam to Angola—caused profound division within our society. And the proposition that to some extent we had to collaborate with our adversary while resisting him found a constituency only with great difficulty; the emotional bias was with the simpler verities of an earlier age. Liberals objected to the premise of irreconcilable conflict and to the necessities of defense; conservatives would not accept that an adversary relation in the nuclear age could contain elements of cooperation. Both rebelled against the concept of permanent exertion to maintain the global balance.[13]

Leaving aside the wisdom of dividing America so neatly into conservatives and liberals, the key concept here is "permanent exertion." Leaving aside, as well, the question of whether this ingredient has been missing from the American world view from the start, Kissinger thought the problem could be remedied. A second Nixon administration, he believed, could educate the American public "in the complexity of the world we would have to manage." But then the headmaster of the school fell from grace, and the professor, even with two more years under Ford, ran out of time. Kissinger acknowledges that "serious men and women whom I respect consider our approach incompatible with the American psyche." Still, he clearly believes that the Nixon approach could have worked, had it not been for Watergate. The record of the three quite different presidents who have succeeded Nixon in the White House does not bear him out.

If there is an American constituency for the sort of "permanent exertion" Kissinger recommends, it has not been proved by the most elemental test: No serious candidate for president has advanced this Kissingerian concept as a prerequisite for sound policy, and still less has any successful candidate claimed it as an electoral mandate. Geopoliticians are free to think in terms of sustained struggle, and diplomats and military strategists may delight in the fine-tuning of deterrence and defense postures. But elected officials have to keep it simple and sweeping. "Peace and prosperity" are the stuff of political campaign commitments.

It is not only the public's attraction to broad formulations that gets in the way of more subtle, nuanced policies. If Watergate and the Vietnam aftershocks can partially explain the disintegration of Kissinger's grand design and the disorderly American world view that carried over into the Ford presidency, the Carter administration was afflicted by disorder of a different sort. In Cyrus Vance, Carter had as his secretary of state a low-key, pragmatic establishmentarian who took a careful lawyer's case-by-case approach. In Zbigniew Brzezinski, his national security adviser, he had a conceptualizing academician with the same taste, if not necessarily the same talent, as Kissinger for sophisticated strategic thinking on a global scale. The irrepressible Brzezinski and the soft-speaking, unflappable Vance were a bad match. Their antagonism was all the more a liability for the country, given Carter's inability to resolve, or even at times to recognize, the guerrilla warfare over policy that was taking place.

Whenever there was a clear consensus within the administration on particular propositions—the Panama Canal treaties, normalization with China, and the Camp David initiative on the Middle East—Carter not only scored significant foreign policy successes but also skillfully rallied public support for his purposes. However, he was unable to develop a consensus within his own administration on a strategy for East-West relations. As a result, he was no more successful than most of his postwar predecessors in rousing the American public out of its natural disengagement to make the "permanent exertion" that Kissinger considered a prerequisite for effective U.S. foreign policy in the nuclear age.

Nowhere was this failure more neatly illustrated during the Carter years than in Africa, an area that has traditionally attracted only sporadic American attention. Indeed, to understnd how the stage was set in the United States for Afghanistan, it is helpful to recall the bitter policy struggle within the Carter administration over what to do about crises in the Horn of Africa and the Shaba province of Zaire in 1977. Much in the way that the Bay of Pigs and the Cuban menace were laid at the doorstep of the Kennedy administration (the ill-fated invasion had in fact been conceived by the Eisenhower administration, on whose watch Cuba had succumbed to Fidel Castro), so were Ethiopia and Angola perceived almost exclusively as Carter "losses." (In fact, both had come under Marxist influence before Carter came to power, and the first Ethiopian-Soviet arms deal was struck in December 1976.)

Nevertheless, as Vance later wrote in his memoirs, "What we did in Africa in the early months of 1977 would have a major effect on Third World perceptions of our policy toward the developing nations, and would set the tone for the remainder of the administration." Vance was leery of plunging into an old and complicated conflict between Somalia and Ethiopia over the Ogaden desert. For a time, the Soviets were working both sides of the street, with a naval facility at Berbera in Somalia and a new role as an arms supplier to Ethiopia. Vance wanted the United States to use similar tactics, retaining some influence in Ethiopia while improving relations with Somalia. He thought flat-out opposition to Ethiopia (where leftist officers had overthrown America's old friend, Emperor Haile Selassie) and unconditional support for Somalia would only raise the risk of an East-West showdown where none was necessary or appropriate.

Brzezinski, on the other hand, wanted to checkmate the Soviets. At one point, over the objections of both Vance and Defense Secretary Harold Brown, he advocated a deployment of U.S. naval forces as a show of power. Vance envisaged little public or congressional support for gunboat diplomacy and, further, feared that if the Ethiopians prevailed, the United States would be seen needlessly to have suffered a defeat. Taking a longer view, he argued that in time the Soviet connection would sour in Ethiopia, just as it had elsewhere in Africa and the Middle East, and Ethiopia would turn again to the West. In the event, the Soviets opted for Ethiopia, successfully exploiting resentment of the U.S. ties to Haile Selassie, and America inherited a new friendship with Somalia without having become involved militarily.

The same philosophical and strategic differences came into play in the complex crisis involving an attack on the Shaba (formerly Katanga) province of Zaire by dissidents operating out of Angola. Vance tried to find a diplomatic solution consistent with the concerns of America's friends and allies in Africa and Europe. Again, the emphasis from Brzezinski's National Security Council staff was on East-West confrontation, with the attackers seen as agents of the Marxists in Angola and the regime of Mobutu Sese Seko in Zaire as a beacon of Western interests, if not values. (In fact, the invaders were remnants of the old secessionist administration in Katanga, which opposed any central government in Zaire; they had no special affinity for the left-wing government in Angola.) The Vance approach prevailed once more, largely because of congressional and public sentiment against any U.S. military involvement in Africa, especially on the side of the rather unsavory Mobutu regime in Zaire. But the argument did not end with these decisions. What Vance thought was successful U.S. preventive diplomacy was anonymously portrayed by those whose advice had been rejected as a demonstration of American weakness. "The lurching back and forth in public about linkage, about the significance of Soviet and Cuban activities in Africa, and the impact of the Shaba and the Horn on U.S.-Soviet relations . . . was hurting the president politically," Vance argues. "It was also undercutting our ability to conduct a consistent and coherent foreign policy."

In October 1978, after an extensive high-level policy review, Carter rejected the idea of linking Soviet and Cuban activities in Africa to U.S.-Soviet relations in general and to arms control negotiations in particular. The point is not so much whether Brzezinski or Vance had the right approach as it is that the tension between them created a perception of disorder not only in the rest of the world but also on the domestic political front. The dispute suggests one more reason why disengagement is the natural American condition. Engagement in far-flung places like central Africa (or, indeed, Vietnam) is difficult to carry off neatly and effectively in the open society bequeathed by Adams and his contemporaries. It is not even easy when the engagement involves old friends with shared democratic values and with whom the United States has the closest of connections, as witness the uncertain American slide into World War II, preceded by deep divisions in Congress over military conscription and lend-lease.

It took Pearl Harbor to produce a national consensus, and then the habit of bipartisanship carried over into the immediate postwar years in the face of the growing menace of communism. A bipartisan foreign policy establishment had been created in the war years and for a time, for reasons having to do with the way Congress was organized before it reformed itself and robbed itself of most of its capacity for strong leadership, the United States engaged in great international enterprises that one would have to say went well beyond the bounds of the Adams Doctrine. A supremely powerful America felt comfortable stationing troops in Europe and rimming the Soviet empire with pacts committing the United States to come to the defense of countries big and small.

But when it came to actually doing something about these commitments, Harry Truman chose the cover of a United Nations "police action" for his commitment of American forces to Korea rather than the more natural course of asking Congress to declare war. And the duplicities of Lyndon Johnson and then Richard Nixon in Vietnam shattered whatever was left of the postwar bipartisanship. In the early 1970s under the sponsorship of Senator Mike Mansfield, and in the 1980s under the sponsorship of Senator Sam Nunn, the natural instinct to disengage flared up again; such efforts to draw down U.S. forces in Europe undermine the effectiveness of U.S. deterrence, even when the efforts fall short. Dissent of any kind, whether between the legislative and executive branches or within the executive branch, gets in the way of the effective projection of American force and influence because it creates doubt about American endurance.

Ronald Reagan, then, only followed the practice of many of his predecessors who, once they got themselves into trouble, had a habit of pleading for bipartisan support. On April 6, 1984, in a speech at Georgetown University's Center for Strategic and International Studies, Reagan said, "We must restore America's honorable tradition of partisan politics stopping at the water's edge, Republicans and Democrats standing united in patriotism and speaking with one voice as responsible trustees for peace, democracy, individual liberty, and the rule of law." But the rest of what Reagan had to say merely confirms the invalidity of the whole concept of bipartisanship. Far from being an honorable tradition, it is the exception to the rule, an empty exercise, as Reagan demonstrated by the nature of his own appeal; it came accompanied with a charge that congressional "second guessing about whether to keep our men there [in Lebanon] severely undermined our policy."[14] Two days earlier, his outreach for bipartisanship took the form of an even nastier charge, that the effect of congressional debate had been to "stimulate the terrorists and urge them on to further attacks."

In fact, it had taken the Reagan administration a full year after the first landing of the marines in Lebanon to knuckle under to the plain requirements of the War Powers Act and work out an eighteen-month grant of authority to keep the marines in place. A month after this apparent coming together of the legislative and executive branches in a "bipartisanship consensus," a yellow Mercedes truck crashed into the marine compound outside Beirut and killed 241 U.S. servicemen. This is not to say that Reagan wasn't half right; congressional reservations about an ill-defined and mismanaged Lebanese policy undoubtedly sent a signal of slack resolve to adversaries, just as it did, fatally, in Vietnam. But it was Reagan's flouting of the War Powers Act that provoked the congressional dissent. And the congressional dissent, in turn, mirrored a considerable split within the Reagan administration between Secretary of State George Shultz and the Pentagon over the wisdom of having the marines there in the first place.

The recent Lebanese experience, in short, merely reinforces the argument that a forceful, consistent foreign policy is hard enough to make and still harder to execute in an open society. In the free play of partisan, public debate, Americans tend to shoot themselves in the foot in a way that baffles allies as well as adversaries. Afghanistan is an interesting case in point. In Europe, for example, it was seen as an atrocity—but not necessarily a measure of American weakness. Yet that is what Europeans heard Americans saying about themselves, or, rather, Republicans saying about the Democratic administration. The debate over America's weakness, in other words, compounded the appearance of weakness. And yet the history of the Afghanistan affair does not reinforce the conclusion that the Soviets were much affected, in what they did, by concern about how the United States would react; if anything, they were probably surprised to discover that the United States reacted as sharply as it did. For they must have known that there was nothing the United States could do to prevent them from attending to what they considered their vital security interests in Afghanistan. Traditionally a nonaligned buffer state between the Soviet Union and areas of American strategic interest to the south (Iran, Pakistan, and the Persian Gulf), Afghanistan had precariously attempted to retain ties to both East and West. In 1978 the civilian government of President Mohammed Daud showed signs of upsetting that balance by seeking aid from Saudi Arabia, other Islamic countries, and Western nations, in order to ease its dependence on the Soviet Union. When he was violently overthrown and murdered by left-wing army elements in April 1978, there was no hard evidence of Soviet complicity but plenty of reason to believe that the Soviets were pleased with his successor, Nur Mohammed Taraki.

The Carter administration decided that there were no grounds to regard the upheaval as anything other than an internal affair. Though Vance now believes the United States should have expressed its concerns "more sharply" at the time of the coup that brought Taraki to power, his advice at the time was that "we not get involved." The next year, concerned about Taraki's independent streak and his inability to control rebellious Islamic elements, the Soviets engineered yet another coup in Kabul. They put a puppet government, under Babrak Karmal, into power in December 1979 and, at the same time, intervened with full military force. While no one can say for sure what their motives were, it is widely understood that the Soviets could not tolerate the prospect of fundamentalist Islamic predominance in Afghanistan, given its proximity to the Soviets' own Central Asian Moslem population.

Brzezinski saw a darker purpose. The invasion of Afghanistan, he argued, was part of a calculated Soviet thrust toward the Persian Gulf. State Department experts insisted there was an easier way for the Soviets to reach the gulf, if that was their goal.

Whatever the case, appearances matter, and in Congress and among the American public the image prevailed of the Soviet Union on the move, shattering precedent, invading and seeking to conquer a neighbor in a way that had no counterpart even in Moscow's treatment of its European satellites. Afghanistan was seized upon in the United States by those who distrusted détente and opposed the SALT II nuclear arms control treaty. It was seen as reason enough to pronounce both dead.

SALT II certainly was dead, in the sense that after Afghanistan there was no hope for its passage by the Senate; in January 1980 Carter arranged with Senate majority leader Robert Byrd to have it tabled indefinitely. Yet the real cause of SALT II's demise, in the view of many experts, was the failure to push it through the Senate in 1979. While this owed something to opposition to the treaty on its merits, there was also a senseless and inexplicable uproar in the summer of 1979 over the "discovery" by certain politicians of a Soviet brigade in Cuba, a brigade whose existence had apparently been known to the U.S. intelligence community for many years. That it suddenly became an issue could hardly have been an accidental breach of security; more likely, it was a way for opponents of SALT II to dramatize the danger of doing business with the Soviet Union. In that context, Afghanistan merely hammered home the point.

That is not the same thing as saying that the invasion of Afghanistan, by introducing a profound chill in U.S.-Soviet relations, marked a fundamental turning point, and still less that it snuffed out détente. Kissinger, as we have seen, held Watergate accountable for the death of détente. But Brzezinski, in an interview with me in June 1980, while still serving as national security adviser, argued that the crisis in the Horn of Africa actually marked the turning point, the beginning of the end of détente. "If we had taken a strong stand on Ethiopia, we would have had SALT II and détente today," he contended. "But we vacillated. Vance wanted to accommodate, then we overreacted on the Soviet brigade, then we tried to please again, and then we got tough again on Afghanistan."

Brzezinski sensed in Vance a personal Vietnam syndrome. Yet he must have known at the time that he spoke that some form or another of détente was still considered by the Carter administration to be a live possibility. Edmund Muskie, who had become secretary of state when Vance resigned in April 1980 over the ill-fated effort to rescue the American hostages in Iran, was then trying to promote accommodation with the Soviets by salvaging the SALT II agreement.

Muskie had met with Soviet Foreign Minister Andrei Gromyko in May 1980. As he described their exchanges in an interview with me several months later, Gromyko offered an intriguing analysis of U.S.-Soviet relations as seen from the Kremlin. It was evident that the Soviets had miscalculated what the U.S. reaction would be to their move in Afghanistan. It was also apparent to Muskie that the Soviets thought they had little to lose in their relations with the United States; they believed SALT II was a dead letter, that the treaty had been done in by the Carter administration in an act of bad faith. The Soviets had also taken note of an increase in U.S. defense spending, part of a concerted NATO buildup committing its members to 3 percent annual increases. Finally, Gromyko thought the brouhaha over the Soviet brigade was a contrivance, calculated to stir up opposition to SALT II.

Muskie found Gromyko's explanation of Soviet views plausible. The secretary also thought that the U.S. response to the Afghanistan invasion, while obviously not sufficiently punitive to bring about a Soviet withdrawal—an impression he thought the administration had mistakenly conveyed—was hurting the Soviets. Sensing a certain moderation in the tone and content of Gromyko's message, Muskie sought to reestablish U.S. credibility with the Soviets. As a longtime senator, he told Gromyko, who obviously believed that the administration could get the treaty through the Senate if it really wanted to, a little about how the process really worked. He promised that Carter would campaign vigorously for SALT II in the hope of creating a public mandate for its ratification if he won reelection. This Carter did, and Muskie spoke out strongly for the treaty during the campaign as well. Soviet ambassador Anatoly Dobrynin even found an occasion to convey to Muskie that he had noticed the administration's effort and regarded it as "positive."

All this is by way of citing vital signs of life in the concept of détente even after Afghanistan, at a time when U.S.-Soviet relations appeared to be frozen. It is impossible to know how far the process might have been advanced had Carter won. But if there was a turning point in U.S.-Soviet relations in 1980, it came not as a consequence of Afghanistan but at the hands of American voters. For Reagan was elected in a landslide with a supposed mandate, derived from his campaign speeches and the Republican platform, to put negotiations with the Soviets on hold until the United States had improved its bargaining position through a massive defense buildup. Reagan dismissed SALT II as "fatally flawed" and promised a hard-line U.S. foreign policy across the board.

Thus, in the early months of the Reagan administration, the United States was talking tough: "drawing lines in the dust" and "going to the source" of communist subversion in Central America; pursuing a "strategic consensus" against the communist menace in the Middle East. Later there would be talk about the Soviets' "evil empire" and of consigning communism to the "ash heap of history."

It was American manifest destiny reincarnate. But there was precious little evidence for such a popular mandate. After the 1980 election, William Schneider published an extensive analysis of its implications. In sharp contradiction to the foreign-policy-mandate theory of the far right, the economy was found to be the dominant issue of the 1980 campaign for roughly two-thirds of the electorate. On the basis of a compilation of opinion data by a major polling organization, Schneider concluded that foreign policy "was much less salient [than the economy]; the percentage mentioning war and peace, relations with the Soviet Union, the hostage crisis in Iran, and other foreign policy problems ranged between 14 and 19" percent. Reagan had a 6-point advantage in preelection approval ratings in the handling of domestic matters, while Carter actually held a 6-point edge in foreign relations. But Carter's advantage reflected a public schizophrenia in its approach to "peace and strength." While Carter led Reagan by a huge 25 points as "the best candidate for keeping the United States out of war," Reagan had roughly the same lead as "the best candidate for strengthening the national defense," as if the two virtues bore no relation to each other.

Another important, albeit different, reading of the public mood came in 1982, when Reagan lost his working majority of Republicans and conservative Democrats in the House of Representatives. Leaving aside the degree to which this represented a referendum on Reagan (or his foreign policy), the consequence of that congressional election, coming at a time when Reagan's prestige and popularity were high, was growing resistance to his policies in Central America and to the deployment of the marines in Lebanon. Yet another indicator of the mood was the absence of public outcry when, after placing enormous geopolitical stakes on the table in Lebanon, Reagan then folded his hand. The affair in Lebanon and the invasion of Grenada nicely demonstrate the Adams Doctrine at work. There was no stomach for an open-ended and seemingly purposeless U.S. commitment of military force in the Middle East. But there was a surge of public pride in the brief and apparently successful American feat-of-arms in the Caribbean. Again, as with the *Mayaguez*, the "toughest kid" was looking singularly untough in the Middle East. Grenada was popular because the American side won, it was over before anybody had a chance to examine how clumsily it had been carried out, and the gains, such as they were, clearly outweighed the risks.

Yet it is unlikely that Grenada figured in any significant way in the Reagan landslide in 1984. On the contrary, the Reagan foreign policy theme for 1984 was most notable for its significantly tempered anti-communist rhetoric, for the Reagan meeting with Gromyko in mid-campaign, and for the president's considerable effort to convince the public of his dedication to the pursuit of nuclear arms control agreements with the Soviets, and to peace. Even so, in the massive Reagan victory in 1984 there is not much evidence that this aspect of his campaign had an important effect. Once again, as in 1980, economic well-being and the president's personal appeal —the "leadership" factor—outweighed foreign policy by a wide margin in the samplings of voters' concerns.

The question in Reagan's second term, as a lame duck president, is whether the domestic concerns that swept him back into office will be his principal preoccupation as he contemplates his place in history. The conventional wisdom has it that Reagan will wish to leave his impress upon the world, however much satisfaction he may take from the success he has had in establishing the conservative ideological agenda at home. Woodrow Wilson wished for a world "safe for democracy," but when he tried to lead the United States into the League of Nations, the Senate, as they say, did not want to get involved. John F. Kennedy wanted a world "safe for diversity," but did not have the time even to demonstrate exactly what he had in mind. Nixon dreamed of building "structures for peace" for a hundred years and wound up threatening the system in order to save it. Carter's foreign policy keynote address at Notre Dame had echoes of Adams: "We are confident that democracy's example will be compelling, and so we seek to bring that example closer to those from whom in the past few years we have been separated and who are not yet convinced about the advantages of our kind of life."[15] He wanted to make America's values—human rights—the cornerstone of foreign policy. He was the peacemaker, normalizing relations with China, settling a potentially explosive, seventy-year-old conflict with Panama over the canal, negotiating SALT II, and capping it all with Camp David—and then winding up taking the rap for the hostage crisis (the toughest kid was being pushed around) and for Afghanistan (the toughest kid was perceived as a pushover).

In the search for his own legacy in world affairs, Reagan will have to abandon many of the ideological frenzies that characterized his world view well into his first term. His rhetoric was said to have been responsible for a revival of patriotic pride, but there is little reason to believe that it stimulated an American appetite for overseas adventure or even the taking of large initiatives—other than, of course, initiatives for arms control and a general easing of tension with the Soviets. But here, as always, Reagan will be at the mercy of a Soviet willingness to accommodate their own sense of a safe and balanced reduction of nuclear weapons to his. He will be at the mercy, as well, of events—of Soviet behavior in, say, Central America or the Middle East. Perhaps his brightest prospect for historic monuments to his world statecraft would be some progress in the establishment of a coherent, grand design for the management of East-West relations, something of what Kissinger had in mind when he spoke of "linkage" and of "peace as a process."

But Kissinger himself is most eloquent on the limits imposed by the American temperament and the democratic process on the building of enduring grand designs. American pragmatism, he wrote in the first volume of his memoirs, *White House Years,*

> produces a penchant for examining issues separately: to solve problems on their merits, without a sense of time or context or of the seamless web of reality. . . . Yet in foreign policy there is no escaping the need for an integrating conceptual framework. . . . The most important initiatives require painstaking preparation; the results may take months or years to emerge.

Success, he said, requires a "sense of history, an understanding of manifold forces not within our control, and a broad view of the fabric of events."[16] In his second volume, *Years of Upheaval,* he spoke of an American tradition of looking at foreign policy in terms of "discrete self-contained problems each of which could be dealt with by the application of common sense and a commitment of resources. The image has been of an essentially benign world whose harmony was interrupted occasionally by crises that were aberrations from the norm."[17]

That is essentially how Americans have in fact looked upon various international crises, even in the activist immediate postwar years of the Berlin blockade, the landing of the marines in Lebanon in 1958 (Eisenhower's Grenada), and on through the Soviet missiles in Cuba, the intervention in the Dominican Republic, and at least the early, relatively discreet American efforts in Vietnam. From his vantage point, Kissinger looked back on this traditional American view of the world as a crippling liability, deploring it even while accepting it as a fact of life.

From his own, earlier vantage point, John Quincy Adams saw not a tradition obviously, but an element in the national character and temperament, owing in some measure to nothing more than geography, that not only would but should condition the American approach to the world. The point is not that the Adams Doctrine is a perfect prescription for U.S. foreign policy in today's profoundly more complicated and dangerous world. It is that Kissinger and Adams are both talking about an enduring fact of life that contemporary American statesmen will ignore at their peril. American foreign policy has been at its best in contemporary times when it has taken these traditional, natural American perceptions and inclinations into account, and at its worst when it has not.

10 A Quest for Invulnerability

JAMES CHACE

James Chace, an editor of the *New York Times Book Review*, is the author of *Endless War: How We Got Involved in Central America—And What Can Be Done*, *Solvency: The Price of Survival*, and *A World Elsewhere: The New American Foreign Policy*, as well as numerous essays on foreign policy issues. Formerly managing editor of *Foreign Affairs*, he has taught at Yale, Columbia, and Georgetown universities.

THE UNITED STATES has never been isolationist. From the moment the American Revolution was successful, the new nation employed a foreign policy of expansion and intervention. From 1798 to 1945 American governments not only tried to secure the country's own continental borders, but also intervened abroad more than 150 times, in places ranging from Mexico, Central America, and the Caribbean to Europe, the Philippines, and Asia. Relatively few American leaders, whatever their public pronouncements to the contrary, have been able to resist the temptation to meddle in the affairs of other nations.

It would, however, be misleading to confuse this policy of interventionism with that of expansionism, and to attribute both to the power of "manifest destiny." That uniquely Western idea, embodying the belief in a special mission to bring enlightened democracy to other peoples and other nations (regardless of whether they were capable of absorbing it), had a special American variation, providing the rationale first for continental conquests and then for intervention in the Pacific. Eventually, manifest destiny became an archaic concept, and the aggressive expansionism that it justified came to an end. But the American policy of interventionism survived them both, and it persists to such a degree that one of the most heated national debates in the United States today centers on whether or not American military intervention can achieve foreign policy goals.

If American interventionism cannot be said to have its roots chiefly in manifest destiny and the messianic desire to spread liberal democracy and its benefits, can it instead be explained by the desire of American business interests to exploit the resources of various regions? Probably not, given that some of America's most fabled interventions (such as those dating back 150 years in Central America or the long and costly involvement in Indochina) have rarely yielded sufficient returns to justify the costs.

In fact, interventionism as national policy has never been adequately understood, yet its consequences affect American behavior today. The painful legacies of Vietnam and the recent presence in

Lebanon demonstrate this: Despite the sacrifice of American lives, the reasons behind the interventions remain cloudy. But examination of one largely neglected strand in American history can provide a more sophisticated understanding not only of why the United States has always been so quick to intervene, but also why Americans in the 1980s view a complex world with such apparent anguish and anger. That strand is made up, first, of the long but fruitless American search for invulnerability—defined as complete immunity to foreign threats—and, second, of the exaggerated sense of vulnerability that has resulted from that search.

Americans have always sought invulnerability. Whether through doctrines, from the Monroe to the Truman, through military systems, from the great fleet of the late nineteenth century to Ronald Reagan's Strategic Defense Initiative (the so-called Star Wars program), or through simple reliance on geography, American leaders have been tireless in their efforts to obtain a level of security that would be absolute. Because Americans have believed that they could attain invulnerability, nearly every threat to U.S. security has been read as dangerous to U.S. vital interests, even those that in time proved insignificant or illusory. This has been as true during periods of national strength as during times of relative weakness. Indeed, even when U.S. political, economic, and military might were at their zenith, Americans were fearful that power would be stripped from them. Archibald MacLeish, writing at the end of the 1940s, observed that "Never in the history of the world was one people as completely dominated, intellectually and morally, by another as the people of the United States by the people of Russia in the four years from 1946 to 1949. American foreign policy was the mirror image of Russian foreign policy: whatever the Russians did, we did in reverse." Ironically, MacLeish noted, "all this took place not in a time of national weakness or decay but precisely at the moment when the United States . . . had reached the highest point of world power ever achieved by a single state."[1]

This inordinate preoccupation with national vulnerability has been a strong force throughout American political history. The United States may well be, as Thomas Jefferson said, a nation "separated by nature and a wide ocean from the exterminating havoc of one quarter of the globe"; but that has never freed American leaders (as it did not free Jefferson) from the fear that such "exterminating havoc" might creep across the oceans and taint the American nation and character.[2] Jefferson's fears were caused by his distrust of the European balance-of-power system, which he felt was responsible for the savage wars that raged across Europe during the sixteenth, seventeenth, and eighteenth centuries. Subsequent generations of American statesmen shared Jefferson's anxieties about the threatening, infectious nature of alien ideologies and political systems, but redefined and broadened the fears as the international political situation evolved and advancing technologies made any "wide ocean" insignificant. In this way, American concern over vulnerability increased in both degree and international scope, and the range of U.S. intervention grew proportionately.

Nothing demonstrates this better than the experience of Vietnam. For most of two decades—from 1955 to 1974—keeping small and impoverished South Vietnam a noncommunist state was officially regarded as vital to the preservation of U.S. security. The failure to prevail in Vietnam, though not accompanied by any dramatic reduction of American power worldwide, heightened America's sense of vulnerability. The shift away from interventionism that many predicted would be the most lasting consequence of the Vietnam War proved to be only temporary.

The course of events during the 1970s and the first half of the 1980s confirms this. Accompanying the U.S. foreign policy defeat in Vietnam was the oil embargo imposed by the Arab oil-producing cartel in the wake of the 1973 Arab-Israeli war, which demonstrated the vulnerability of U.S. and allied vital resources. The Soviet airlift of Cuban troops into southern Africa in 1975 to aid the new Marxist government of Angola seemed to many Americans to pose a potential threat to sea-lanes in the southern Atlantic. Most dramatic of all, on November 4, 1979, the American people saw Iranian militants seize the U.S. embassy in Tehran and take sixty-six American citizens hostage. In exchange for the release of these hostages, fifty-two of whom were destined to be held for more than fourteen months, the militants demanded the return of the deposed Shah (then receiving medical treatment in the United States) and his financial assets, as well as an apology from the American government for wrongs committed against the Iranian people during the often brutal years of the Shah's reign. It was a signal moment in U.S. history: On the other side of the world, Islamic extremists in a country whose power was no match for that of the United States nonetheless placed the American people in a state of outraged helplessness.

Predictably, the American public grew bitter, and President Jimmy Carter sensed this mood. Increasingly, Carter and his national security adviser, Zbigniew Brzezinski, came to feel that a tangible demonstration of American power and resolve was needed. In April 1980, over the protests of Secretary of State Cyrus Vance, Carter authorized a military attempt to rescue the American hostages in Iran. The result was a disaster, and, worse, a disaster caused not by resistance but by ineptitude. The mission never got past the Iranian desert, where, first, two of its eight helicopters broke down, and then eight men were killed when another helicopter slammed into a transport plane during a hasty pullout. In the aftermath of this incident, the question for many was no longer whether Washington would protect its interests overseas, but whether it could protect them. If America's massive military power could not be mustered with sufficient skill to carry out a political objective, there seemed little left to Washington but the shell of diplomacy.

By the end of the Carter administration, allies and enemies alike had become contemptuous of the United States, complaining that its foreign policy was crippled by incoherence and contradictions.[3] Carter had swung from an accommodating to a hard line toward the Soviet Union, guided at first by the more modulated diplomatic approaches of Vance and later by the mercurial and combative attitudes of Brzezinski. When, in June 1978, Carter warned the American people against "excessive swings" in attitude toward the Russians "from an exaggerated sense of compatibility . . . to open expressions of hostility," his cautionary words could be read as self-criticism. His foreign policy lacked balance and measure.[4]

In fact, such swings in diplomatic attitude had been characteristic of American foreign policy throughout the 1970s. Taken as a whole, these shifts offer compelling evidence of America's preoccupation with vulnerability and of confusion over precisely what constitutes a threat to American security. The sharp American response to the seizing of the U.S. merchant ship *Mayaguez* by Cambodia in 1975, for instance, was orchestrated by the same team —President Gerald Ford and Secretary of State Henry Kissinger— that was to prove unable to affect Angola's civil war and its subsequent swing into the Soviet-Cuban camp. Yet the latter event, placed in historical perspective, was certainly the more important of the two.

The image of American instability and ineffectiveness knew no party lines in the 1970s, and the Iran hostage crisis now seems less an isolated incident than the conclusion of a long and painful process. Still, things were to get worse. In December of 1979, the Soviet Union, after an era of adventurism whose term ran disturbingly parallel to that of the period of U.S. decline, deposed the Afghan president and installed their own man, Babrak Karmal, in Kabul. Babrak's authority was reinforced directly by Soviet land and air forces. Once again, both the U.S. government and the American people were stunned; yet Carter could do little more than institute ineffective economic sanctions against the Russians, shelve the SALT II arms control treaty, and boycott the 1980 Moscow Olympics, none of which seriously affected Soviet behavior.

Asia and the Middle East were not the only regions where American prestige was to suffer during the Carter years. The 1979 decision by the NATO nations to deploy medium-range cruise and Pershing II missiles on European soil in order to reduce the threat of Russia's SS-20 missiles caused an uproar in nearly every society of Western Europe. In what appeared to many Americans to be an ungrateful display of self-interest, large groups of European demonstrators bitterly denounced the NATO decision as a U.S. manipulation to ensure that Europe, rather than the continental United States, would be the first, and perhaps the only, theater of a limited nuclear war. Friends as well as foes, it seemed, were less willing than ever to follow the lead of the United States (although the missiles were eventually deployed more or less on schedule and more or less peacefully).

It was the stated aim of the Reagan administration to end a "decade of neglect," which included the string of episodes that had humbled American prestige.[5] The opening years of Reagan's presidency were characterized by a dramatic rise in every level of defense spending, as well as an increased willingness to speak openly of using such military might. Yet the precise objectives of all this increased spending and blustery rhetoric remained unclear. Presumably they were designed to assuage the feelings of uneasiness and vulnerability brought about by the actions of nations whose size and power had never been, in happier times, sufficient to cause any widespread public concern in the United States. But, as Stanley Hoffmann observed early in 1982, "More of everything is not a policy."[6] Exactly how, then, was the new glut of weapons to improve America's position in the world?

The Reagan administration eventually answered this question

—in Central America, where U.S. support was given to almost anyone willing to fight the "Marxist-Leninist" government of Nicaragua and the leftist guerrillas in El Salvador; in Lebanon, where American marines were deployed to "keep the peace" in Beirut; and on the Caribbean island of Grenada, where the lives of American medical students and the liberties of Grenadians were "saved" from socialist tyranny by a large-scale American invasion. For a time, it seemed that the jingoistic rhetoric of the Reagan administration might be put into practice; it seemed, ever so briefly, that the age of American preeminence might return.

It had become common, particularly among those who grew up during World War II and the immediate postwar period, to view the decline of America's position in the world during the 1970s as an aberration. Central to this view is the belief that the strength of the United States immediately after the defeat of Germany and Japan and before the rise of a nuclear-armed Soviet Union was somehow "natural," a right and proper preeminence that derived from the failure of totalitarianism and realized the long-cherished American dream of global invulnerability. The United States, in 1945, had come into its own: Colonies looked to it to support their demands for independence; Western Europe and Japan needed American help in rebuilding their cities and their nations along democratic and capitalist lines. If the Soviet Union did represent a threat to America's vision of world order, it did not, however, possess the military might to threaten seriously that vision.

It is not difficult to understand why Americans feel a certain nostalgia for the postwar world. But in truth, there was very little about the position of the United States in that world that was "natural." Indeed, if anything in the modern American experience could be viewed as aberrant, it was that postwar preeminence. For although the profound sense of vulnerability that had haunted the Founding Fathers and presidents throughout the nineteenth and twentieth centuries seemed at last to have disappeared, it was in reality only masked beneath a national chauvinism that approached bravado. This bravado was comparatively short-lived—disappearing, as it did, with the appearance of a nuclear-armed Soviet Union—and vulnerability returned to dominate American international behavior. It continues to do so today. Thus it is a feeling far more central to the American experience than any sense of global predominance.

Perhaps the most important reason for these feelings of vulnerability, and a resultant preoccupation with threats from foreign powers, has been the belief that the special quality of the American nation and society—the result of enlightened government, an extraordinary geographical position, and a wealth of natural resources—made the United States an obvious target for all of the world's aggressive or subversive forces. This single consideration has generally overridden, or at least influenced, all other foreign policy discussions. The Monroe Doctrine of 1823 may have been made possible by the protection of the British fleet, and it may also have been designed originally to protect U.S. shores against French, Russian and Spanish power, but the fact remains that before long, American governments were most concerned with the danger of British power. Even though Britain was a liberal, parliamentary state, its imperial ambitions seemed to threaten not only American territorial integrity, but also the American democratizing mission. Through much of the nineteenth century, Americans suspected the British of having expansionist aims throughout North and Latin America. No area aroused more American concern than Central America, where the British held Belize (a coastal enclave alongside Guatemala that was also known as British Honduras), the Bay Islands, and the Atlantic coastal strip of Nicaragua inhabited by the Miskito Indians. A careful reading of history reveals that Britain's activities in the region were intended merely to expand trade; but the possibility of any increased British influence was enough to alarm an America that still had vivid memories of the War of 1812.

By 1860 the threat of actual conflict with the United States had caused the British to withdraw from most of their Central American holdings. Yet for the next half-century, a period when the United States expanded its own political and economic influence over Central America, Washington frequently voiced fears of British encroachment in the region. Finally, this preoccupation with vulnerability in Latin America seemed to abate after Britain accepted American arbitration in 1895 in a boundary dispute between itself (in British Guiana) and Venezuela. On the occasion of the successful resolution of the matter, President Cleveland's secretary of state, Richard Olney, declared that "today, the United States is practically sovereign on this continent, and its fiat is law upon the subjects to which it confines its interposition."[7]

Undoubtedly this was true, but what the secretary failed to note was that this had been the case for half a century. By eagerly embracing the maritime strategy of Captain Alfred Thayer Mahan, which envisioned America as having a new global reach, American policymakers had tried to ensure U.S. invulnerability against any real or supposed threats to the nation. It is important to remember that so-called American isolationism was confined solely to the relationship with Europe, not with the Western Hemisphere. Not only did Washington intervene time and again in Central America and the Caribbean, but it also remained on the lookout for any possible external threat. For example, the socialist turn taken by the Mexican revolutionary government from 1915 to 1920 was believed to be threatening to American security; this time the enemy was Bolshevik Russia instead of the British Empire. But just as had been the case with the British, there was little persuasive evidence of any actual threat to U.S. security beneath all the talk of the supposed ambitions of a rival power.

At the end of World War II, however, even the imagined threats to U.S. security all but disappeared, and Americans should have been able to enjoy the relative calm that global predominance offered. But the age-old feeling of vulnerability never really went away. All that was needed was a candidate to fill the role of antagonist for an America that still believed, as it had since the days of the Founding Fathers, that the special nature of American society —both material and moral—made it an irresistible target for aggression and subversion. With the Soviet development of the atomic bomb in 1949, the hydrogen bomb in 1953, and finally a full force of ballistic missiles during the 1950s and early 1960s, American fears were amply fulfilled. By the mid-1960s America was genuinely vulnerable again, first and foremost to Soviet missiles, but also to an apparent incapacity to act decisively and effectively in Indochina, where it seemed impossible to preserve a noncommunist South Vietnam, despite having sent half a million troops and spent well over $150 billion.

The United States had to acknowledge not only the Soviets' new nuclear parity or, as it was sometimes called, "essential equivalence," but also Moscow's growing ability to project power throughout the globe by using conventional forces. The Soviet Union was initially perceived primarily as a continental power. While its sway over Eastern Europe was assured, its one excursion into Cuba in the early 1960s seemed to demonstrate that when it tried to act so far from its shores, Moscow was dangerously overextended. But in the 1970s the Soviets were able to intervene in southern Africa by airlifting Cuban troops and equipment into Angola. Subsequently, the Soviet Union became involved in the Horn of Africa, where it switched support from Somalia to Ethiopia and then helped Ethiopia turn back an invasion from Somalia and cope with a guerrilla war in Eritrea.

This new demonstration of Soviet power further alarmed Washington, which feared the Soviets could, if they chose, threaten the Persian Gulf, a region thought to be vital to the United States and its allies in Europe and Japan. The Ford and Carter administrations were unable to flex military muscles in return, in part because of the 1973 War Powers Act, which restricted the right of the president to commit U.S. forces overseas without explicit congressional approval. Bristling under such restraints on executive action abroad, a clear legacy of Vietnam, Secretary of State Henry Kissinger inveighed against Congress, complaining that "America seems bent on eroding its influence and destroying its achievements in world affairs."[8] In fact, what most characterized American foreign policy in the late Kissinger era were bluster and bluff.

Nonetheless, the achievements of the Kissinger years in dealing with the Soviets were not matched by the Carter administration. The conclusion of SALT I in 1972 and the lineaments of a second arms control agreement at Vladivostok in 1974 were not built upon by Carter. Instead of turning the Vladivostok agreements between Ford and Brezhnev into a second SALT treaty, Carter sent radical new proposals for arms cuts to Moscow after just a few months in office. Predictably, they were rejected. By the time the matter was straightened out and a SALT II agreement was ready more than two years later, relations between the United States and the Soviet Union had worsened in other spheres. Finally, with the Soviet invasion of Afghanistan in 1979, there was little prospect for ratification of SALT II by Congress, even if the president had been disposed to ask for it.

While America found itself unable to halt Soviet adventurism and, in the eyes of conservatives, had become vulnerable to the destruction of its land-based missile systems by a Soviet nuclear first strike, it also found itself dangerously vulnerable on the economic front. This too was in part a legacy of Vietnam, when Lyndon Johnson refused to increase taxes to pay for the war. LBJ's policy was to fight a costly war without inflicting economic pain on the public or disturbing the promise of a consumer society. The resulting deficit—reflecting the amount the consumer and the government spend in excess of what the country earns through the production of goods and services—created inflation, as the government simply printed more currency to make up for the shortfall. (This contrasts with the situation in the mid-1980s, when foreign capital, especially from Europe, helped finance the huge fiscal deficit.) Rising prices for domestically produced goods and services meant that the United States sold less abroad, which caused its trade balance to decline. While 1964 had seen the highest American trade surplus since 1947, by 1971 the United States experienced its first trade deficit since 1893.

Since the dollar was also the key international currency, providing the liquidity for global trade, U.S. indebtedness meant that the Treasury no longer possessed enough gold to back the dollars that had been printed and sent abroad. Foreigners were expected to continue using dollars as world money, even though there was no gold behind them. In 1971, when Nixon announced that the United States was no longer willing to convert dollars into gold, foreigners' eagerness to hold U.S. dollars—their faith, that is, in the American economy—declined sharply.

Like its predecessors, the Kennedy and Johnson administrations, the Nixon administration interpreted these deficits as the inevitable result of America's global commitment to maintain the peace. This commitment had been undertaken after World War II and soon reached unnatural proportions; the United States could not afford to maintain it in a more economically competitive world. Yet this predictable development was interpreted by most American policymakers as "temporary," and a profound sense of vulnerability accompanied it.

For most of the 1970s, then, the United States financed its external trade deficit, which went from $6.4 billion in 1972 to $34 billion in 1978, by simply printing more money. Despite rising inflation, the Carter administration went ahead with a program of domestic economic expansion, which contributed to a further increase in the inflation rate, from 11.3 percent in 1979 to 13.5 percent in 1980.[9]

By 1978, for the first time in memory, the United States faced a situation where the weakness of its currency on the world's exchanges affected the domestic life of its citizens. The Federal Reserve Board, responsible for the size of the money supply, tried to defend the value of the dollar abroad by raising interest rates at home. This, in turn, adversely affected American consumers and cut business investment. A year later the inflation rate grew worse, the dollar fell further, and the Federal Reserve again raised interest rates, this time to unprecedented levels. As might have been expected, with the shrinking of the money supply, inflation finally fell. The Federal Reserve under Reagan continued these policies, and in time interest rates came down as well.

Restricting the money supply was the only way the Reagan administration could handle its conflicting commitments to a huge military expansion, tax cuts, and an end to price inflation. But because the administration was committed to restoring American preeminence through a defense buildup and could not cut domestic spending as severely as it wished, record-high fiscal deficits resulted. Instead of printing money, the government under Reagan borrowed on the credit markets. Initially, a severe recession resulted. But during the second half of Reagan's first term, the enormous fiscal deficits brought about a significant recovery in America.[10] An ironically Keynesian policy had worked for an avowedly conservative administration.

In most circumstances a restricted money supply would be unable to finance such a recovery. But the United States was able to use external sources to finance its own domestic imbalance. High U.S. interest rates attracted foreign money, as overseas investors increasingly came to see the dollar as a currency of refuge in a politically troubled world. The inflow of foreign monies offset the high domestic accounts deficits, while pushing the dollar to new highs on the foreign exchange markets. By compensating for the drain on domestic savings caused by the federal government's deficits, foreign capital paradoxically permitted the world's richest country to become its biggest borrower.

Although interest rates did fall during the economic recovery that accompanied the end of Reagan's first term, they were still high by any postwar standard. Because the inflation rate had also dropped, the point spread between the rate of inflation and the rate of interest continued to make foreign investment very attractive. Much of it, however, was in short-term financing or in Treasury bills that the investor could easily dispose of and did not represent a significant investment in the production sector of the American economy. At home, however, the lower rates encouraged domestic borrowing and investment, and stimulated the recovery.

Along with the size and diversity of the U.S. capital market, the administration also offered higher after-tax returns on capital. A dynamic American economy ushered in a second Reagan term.

Though the administration has denied it, America's huge deficits can ultimately be covered only by printing money with abandon or borrowing on a scale so immense as to keep interest rates unacceptably high. In either case, the American economy may head into a new crisis: Printing money would set off a new round of inflation and cause foreigners to withdraw their capital with sudden urgency (causing a rapid fall in the dollar's value), or the Treasury would have to borrow at such high rates from domestic credit markets that the resulting credit squeeze would probably produce a new recession.[11] Aggravating this condition is the continuing trade deficit, which amounted to $123.3 billion in 1984. As the world buys fewer overpriced American goods, U.S. industry will inevitably slow down. As long as its deficit mounts, the United States is seen as a country desperate to borrow: In 1985 it is likely to become a debtor nation for the first time since 1919. Before long it could well replace Brazil as the leading debtor nation in the world. (In the meantime, Japan is moving in the opposite direction and becoming the world's leading creditor nation.) Other people flock to buy dollars, but not, as in the heyday of the American economy in the 1950s, American goods.

At the same time as the highly valued dollar made foreign goods cheaper in the United States, there were new efforts to enact protectionist trade measures, aimed especially at the Japanese challenge. In 1984, for example, U.S. total merchandise imports amounted to a huge $341.2 billion. The 1980s were indeed a far cry from the carefree days of the 1950s, and also from the early 1960s, when inflation was running at 1 to 1.6 percent[12] and interest rates were at a comfortable 4.5 to 5 percent.[13] Then, with the Vietnam expedition barely under way, America's commitments and capabilities were roughly in balance and a youthful president called for a new assertion of American power. But Kennedy's vision was of a limitless future, a vision that would prove as fleeting as the postwar world in which it was born.

By the end of its first term, the Reagan administration had redefined the terms of America's vulnerability. The new definition harked back to the 1947 statements of Dean Acheson, in which the then undersecretary of state explained how "Soviet penetration" of the Balkans and Greece would eventually "infect" the Middle East and Europe, "like apples in a barrel infected by one rotten one."[14] Reagan's view also recalled the Truman Doctrine, which, with its emphasis on the domino theory, sought to defend Eastern Europe and the Mediterranean. Indeed, the Reagan administration expounded the domino theory with a gusto not heard since the escalation of the Vietnam War. Secretary of Defense Caspar Weinberger, when asked which areas of the world had highest priority in the effort to reassert America's position, answered, "All of them."[15] Which is to say, there are no priorities, even though, as Theodore Draper has written, "unlimited commitments required unlimited power."[16] When pressed, administration officials have focused on Central America and the Middle East as places where America's vital interests are most threatened.

The administration has been especially keen on labeling Central America as vital to U.S. security. The small nations of that region are seen as quintessential dominoes. In a speech to a joint session of Congress on April 27, 1983, Reagan dramatized their proximity to the United States—the final domino. "El Salvador is nearer to Texas than Texas is to Massachusetts," the president said. "Nicaragua is just as close to Miami, San Antonio, San Diego, and Tucson as those cities are to Washington." And "if we cannot defend ourselves [in Central America], we cannot expect to prevail elsewhere. Our credibility would collapse, our alliances would crumble, and the safety of our homeland would be put in jeopardy." In an address before the National Association of Manufacturers, the president again evoked the danger that Central America might fall to communism:

> At stake in the Caribbean and Central America . . . is the United States national security. . . . Soviet military theorists want to destroy our capacity to resupply Western Europe in case of an emergency. They want to tie down our attention and forces on our southern border and so limit our capacity to act in more distant places, such as Europe, the Persian Gulf, the Indian Ocean, the Sea of Japan.

According to this reasoning, the loss of El Salvador to Marxism-Leninism would mean, in the words of William Clark, the former national security adviser, that "Mexico and the United States become the immediate rather than the ultimate targets."[17] The truth is, of course, that if troubled Mexico were to undergo revolutionary change, this would not necessarily come about because of any "targeting" of that nation by the Marxists of Nicaragua or El Salvador. Similarly, the possibility of the United States being threatened by Cuba or the Soviet Union through El Salvador or Nicaragua must be considered unlikely at best. The United States is certainly not vulnerable to the Central American brand of Marxism-Leninism, as represented by the military forces of Nicaragua or the guerrillas in El Salvador. The idea that the Soviet Union would use these tiny countries as a springboard for an assault of its own is so remote as to be unthinkable.

Nevertheless, it is just this kind of broad generalization that often characterizes American foreign policy. In the Reagan era, the problem is unusually severe. Officials skip from domino to domino without ever examining the unique character of each situation. "We are dealing here with one of the most treacherous forms of political thinking," Theodore Draper has written.

> The analogies here are physical ones—apples and dominoes—as if the political world obeyed the most rigorous physical laws. . . . how and why does a rotten apple in Greece make other apples rotten as far away as Iran and all to the East, Africa through Asia Minor and Egypt, Europe through Italy and France? How and why . . . could El Salvador and Nicaragua make Mexico and the United States the targets for the Soviet brand of communism?[18]

Even when falling dominoes do not directly threaten U.S. security, they are assumed to undermine national credibility. So argued the National Bipartisan Commission on Central America, chaired by Kissinger, in its final report, published in January 1984. If Washington should permit the existence of a Marxist regime in Nicaragua, or abandon El Salvador to its own devices, thus risking installation of a Marxist regime there, according to the commission, then the United States would be seen by enemy and ally alike as lacking the power to control its own sphere of influence, with accompanying grave consequences for the "global balance of power." The commission went on to cite the threat to the Panama Canal, and the danger of a flood of refugees, "perhaps millions of them, many of whom would seek entry into the United States." In short, it said, "the crisis is at our doorstep." Then came a telling phrase: "our credibility worldwide is engaged." The "triumph of hostile forces in what the Soviets call the 'strategic rear' of the United States would be read as a sign of U.S. impotence."[19]

But the commission never specified precisely the security threat "at our doorstep." The Soviets installed in Cuba can already threaten U.S. access to crucial sea-lanes, as the commission points out, and Americans have managed to live with this for twenty years. If Mexico were to install a government hostile to the United States and invite foreign military advisers and offensive weapons, that might well constitute a threat to the United States; but even Mexicans do not believe this could happen as a result of events in Central America, but rather because of the country's own social and economic troubles. If so much is really at stake in Central America—from the credibility to the U.S. alliance structure—and if U.S. security is so gravely threatened by the possibility of a Marxist regime in El Salvador, then how could any president not commit American resources to overthrowing the Sandinistas in Nicaragua and destroying the revolutionary left in Salvador? If the stakes are not so high, however, then American foreign policy has once again been distorted by an inordinate preoccupation with vulnerability. In this respect, U.S. "credibility" is indeed threatened, but only because of the obtuseness of American foreign policy.

Before Reagan had withdrawn the last U.S. forces from Lebanon in 1984, he declared that it, too, was "central to our credibility on a global scale"; it, too, was a "vital interest." In fact, it was not Lebanon itself that seemed vital in Reagan's view; rather, it was the behavior of various alien forces in Lebanon, particularly Israel, Syria, the Soviet Union, and the Palestine Liberation Organization. Concern over the ambitions and actions of these forces predated actual U.S. involvement in the multinational peacekeeping force in 1982.

The Reagan administration had one overriding concern in the Middle East when it took office in 1981—to exclude Soviet influence from the region. This implied, of course, the maintenance of a firm commitment to Israel; but it also implied strengthening ties to Arab states, such as Egypt and Saudi Arabia, which showed little inclination to fall into the Soviet camp, as well as supporting groups and policies that would reduce the stature of the Soviet Union's Middle Eastern clients, especially Syria and Iraq. This had meant showing significant sympathy for Israel when its air force destroyed an Iraqi nuclear reactor in the summer of 1981. The Israeli invasion of Lebanon presented the administration with its best hope of attaining its objectives in the Middle East—the defeat of the PLO, the humiliation of Syria, and the subsequent erosion of the Soviet presence.

Despite Reagan's repeated claims that the United States was interested only in seeing "the three foreign factions—the PLO, the Syrians, and the Israelis—out of Lebanon," and his insistence that "we are pushing on that as fast as we can,"[20] subsequent developments, as well as the written statements of various men directly involved at the time, reveal that Washington was far from opposed to Israel's invasion, which began on June 6, 1982. It has even been argued that the administration, by never strongly warning against such an action, even though high officials knew of its possibility in advance, gave the green light to it. Certainly this is the Israeli interpretation.[21]

The Israeli invasion dealt severe blows to the Syrian army and air force, and it drove the PLO out of Beirut. Yet, in the end, the Lebanon experience could hardly be described as a foreign policy victory for the Reagan administration. The United States lost its embassy and the lives of 262 servicemen, ultimately having to withdraw its peacekeeping force without having substantially affected the Lebanese situation. None of the long-standing bitterness between religious factions was eased; the foreign armies in Lebanon showed little inclination to leave; and the Syrians ultimately proved strong enough to recover from their 1982 defeats. Perhaps most critical of all, Middle Eastern nations aligned with the Soviet Union were presented with good reason to strengthen their ties to Moscow, while nonaligned Arab states viewed the U.S. error of omission—its failure to prevent, or at least effectively condemn, the Israeli invasion—as proof that the Reagan administration cared only about America's "special relationship" with Israel.

The Reagan administration's Lebanon policy, in nearly every sense, backfired. Consistent with much of U.S. foreign policy behavior during the 1970s, it had brought defeat on itself. By viewing the problems of Lebanon through the lens of the Cold War—that is, by failing to see it as anything more than an apple or a domino—the administration failed to come to terms with that nation's actual problems, which have more to do with long-standing internecine religious feuds than they do with the global balance of power. But Reagan and his aides never acknowledged this. As the president neatly put it, "If others feel confident that they can intimidate us and our allies in Lebanon, they will become more bold elsewhere."[22] Once again it appears that events in a small, distant, overrun country contribute to the vulnerability of America.

In recounting his experiences as secretary of state, Alexander Haig claims that the U.S. failure in Indochina was due, at least in part, to the fact that America "refused to treat the Vietnam insurgency as anything other than a local problem" and did not "take the issue to the Soviet Union."[23] But, in fact, no American president, from Kennedy to Nixon, viewed the Vietnam issue as primarily a "local problem"; the attempts by Kissinger to negotiate a peace settlement, particularly during the crucial year of 1972, only began to bear fruit when, to use his own words, he stopped "holding Moscow responsible for Hanoi's" policies.[24] Despite their historical inaccuracy, Haig's remarks shed light on the attitude of the Reagan administration toward both Central America and Lebanon. Such areas are seen first and foremost in the context of the global balance of power. Regional problems and characteristics are relegated to secondary importance.

No single issue more clearly demonstrated the confusion and the penchant for self-defeat present in both the Carter and Reagan foreign policies than did the Euromissile debate. When the NATO nations voted in 1979 to deploy 572 intermediate-range missiles (108 Pershing II's and 464 ground-launched cruise missiles), their reasons were manifold. Concern over Soviet deployment of the new SS-20 missiles, which were aimed at Western Europe, and over Soviet development of the "Backfire" bomber caused European leaders to seek the protection of a new "nuclear umbrella" based not in the United States but in Europe itself. In addition, some European leaders, particularly West German Chancellor Helmut Schmidt, believed that once a decision was made to accept these missiles, it would be easier for the United States to make substantive new offers on arms limitation to the Russians. This was the so-called two-track policy, and in Schmidt's mind it was a fair tradeoff. From the first, then, the issue was more a political than a strategic decision; there was no real reason at the time for it to become more than an intra-alliance affair. If the deployment had been matched by substantive arms control negotiations, as Schmidt hoped, the entire issue might never have caused widespread public concern.

But in 1980 the Carter administration had little interest in either sounding a soft line in Europe or accommodating the Soviet Union on arms control. Americans were still being held hostage in Iran; the Russians had invaded Afghanistan; and, with a presidential election approaching, Carter and his aides had no desire to be tied to a policy of negotiating with the Soviets that a growing number of Americans, including conservative Democrats, were suddenly equating with the appeasement of Nazi Germany before World War II. The controversy was further complicated by Carter's vacillation over, and eventual decision not to deploy, the enhanced radiation weapon, or neutron bomb, which was known to be favored by at least some members of the administration. Carter, declaring that the Russian invasion of Afghanistan "could pose the most serious threat to world peace since the Second World War," began to heighten the strategic overtones of the Euromissile debate, thus increasing the fears of the European population both of a Soviet attack and of an uncertain American commitment to arms control measures.

Deploying the missiles soon became more important to U.S. policymakers than it did to most of their European counterparts—even though there was no clear military need for them. Neither the Carter nor the Reagan administration made it clear to the American and European public that emplacement of the missiles was Europe's choice and did not affect the credibility of the U.S. nuclear deterrent. But soon the Soviet Union moved cleverly to exploit fears in Western Europe that the United States either would not protect Europe once these missiles were deployed or, alternatively, would draw Europe into a war by further threatening the U.S.S.R. by their deployment. At that point, deployment became a political necessity for the alliance.

Tension over the two-track policy increased after the Reagan administration took office. By October 1981, two hundred and fifty thousand people marched in West Germany alone to oppose the deployment of the new missiles, and there were vast protests elsewhere in Europe. The antipathy was not directed solely at the United States, but also at the European governments, some of whom—despite having participated in the original deployment decision—now felt caught between the designs of the largest NATO ally and the demands of their own people. Schmidt, for example, was seriously weakened by the uproar, when his own Social Democratic party took a position against the missiles. The opposition never rose to the point where it was actually able to prevent deployment;

indeed, in October 1982 Christian Democrat Helmut Kohl, a strong advocate of the missiles, was elected chancellor of West Germany to replace Schmidt. But the entire experience put a severe strain on both the Atlantic Alliance and U.S.-Soviet relations.

While the deployment of the missiles began on schedule and most of the protests died down, few of the underlying issues and tensions were successfully resolved. Strong antinuclear feeling persists throughout Western Europe, and any new crisis in U.S.-Soviet relations, such as another breakdown in arms control talks like the one in late 1983, is likely to bring forth a new spate of protest in Europe and to deepen distrust of American leadership. By introducing the very dubious issue of nuclear superiority in space, the Reagan administration has increased the possibility of another political crisis over strategic issues. In his Strategic Defense Initiative (SDI), the president has proposed to defend the United States from Soviet missiles through a system of devices operating in space. The plan has already been judged by many scientific and technical authorities to be unfeasible, except on the most rudimentary scale, and incapable of being tested in any meaningful way.

NATO defense ministers, meeting in March 1985, gave perfunctory endorsement to the research phase of SDI, in part because they might like to share in the $26 billion high-tech effort. The British Foreign Minister, Sir Geoffrey Howe, probably best expressed the polite reservations of the Europeans when he warned that the "risks [of SDI] may outweigh the benefits," because the Star Wars program might make deterrence harder, not easier. Should the administration, while continuing research on such a system, refuse to negotiate seriously with the Soviets over the control of space weapons, the standoff is bound to heighten European fears of the result. Any U.S. plan for even a limited defense of its land-based missiles raises concern in Western Europe that its defense will no longer be tied to the defense of the United States. This prospect of possible abandonment by the United States stirs up controversy within Europe over Western defense issues, and before long American conservatives will doubtless begin to accuse the Europeans of losing their nerve.

Certainly, the nuclear defense of Western Europe should not be viewed strictly in the light of U.S.-European or even U.S.-Soviet relations. Much of the recent conflict is the result of Europeans having been willing, since the end of World War II, to grow and prosper under the umbrella of U.S. defense, and then suddenly being faced with the problem of how to take an increased responsibility for that defense themselves.

But given that American moderation—and mediation—could facilitate this process, as well as set a more promising stage for the larger issues of arms control, both the Carter and Reagan administrations have acted in a fashion that raises serious questions throughout the alliance about the priorities of American foreign policy. Presumed U.S. nuclear vulnerability, both administrations concluded, was to be handled by increasing U.S. nuclear strength, although no Soviet action anywhere could clearly be attributed to Soviet nuclear superiority. In 1983, when the White House appointed a Commission on Strategic Forces, chaired by former national security adviser Brent Scowcroft, to examine the capability of U.S. forces, even the legendary "window of vulnerability" was found no longer to exist. Allies as well as adversaries thus have good reason to be perplexed by the signals emanating from Washington. While insisting that the strategic balance and the arms race were vital issues, both the Carter and Reagan administrations have finally subordinated them to geopolitical concerns in the Middle East and Central America. And, while the resumption of U.S.-Soviet arms talks in March 1985 can be seen as an attempt to redress this balance, Reagan was only willing to enter such negotiations, with a broad agenda, after he had spent his first term indulging in a massive buildup of American military might—that is, after his own sense of American vulnerability had been eased.

America's need in foreign policy, as in domestic policy, is to act in accordance with its means, both economic and military, distinguishing its vital from its secondary and general interests. But what are these vital interests? Securing the position of the United States within its own hemisphere, without threatening the independence of others, is surely one. The safety of Western Europe, Japan, and, for many special reasons, Israel remains vital to American interests, because of strong economic ties and shared moral and political values. Americans have long believed that any Soviet expansion into these areas would be a direct threat. Much of the Third World beyond the Western Hemisphere remains a matter of general American interest—such as southern Africa, where a racial war could break out that would be of deep concern to American citizens, or Thailand, which could be affected by the Vietnamese occupation of neighboring Cambodia—but it becomes a more serious interest only insofar as the United States or its allies remain heavily dependent on foreign natural resources (as, for example, Europe and Japan depend on Persian Gulf oil).

By failing to distinguish between vital and general interests, recent U.S. foreign policymakers have often simply linked together various problems and dealt with them through bursts of anti-Sovietism. Yet there are many foreign policy problems that profoundly affect America's vital interests which either do not involve the Soviets directly or actually invite cooperation with them. An example of the first type of problem was the debate on transferring jurisdiction over the Panama Canal to the Panamanians; stripped of its jingoist overtones, that debate involved questions of America's needs within its traditional sphere of influence, but no Soviet threat was at stake. Among the second category of problem, on which U.S.-Soviet cooperation would be appropriate, are issues that threaten the international order, such as the environment, the exploitation of the seabed, and nuclear proliferation.

Although not every Soviet move is directed against the United States, it is understandable, and often necessary, for Americans to worry over Soviet expansionism. Soviet support of Vietnam, for example, is primarily aimed at the containment of China, but there is always a danger that Vietnam, having expanded into Cambodia, might thrust into Thailand, whose government is aligned with the United States. If that were to occur, Americans would face difficult decisions about the nature and extent of their support for Thailand: whether to provide military aid, military advisers, or direct armed support. To answer such questions, it would be necessary to determine the degree of the U.S. commitment, given finite military and economic resources, to confront the Soviets in areas of marginal interest.

Anti-Sovietism is at once too much and not enough to constitute the essence of an effective American foreign policy. What is needed instead is a policy that opposes the Soviets only at points where interests clearly collide (and where the United States has the resources to oppose them effectively), while allowing room for cooperation on issues of mutual concern. Beyond this, the United States needs a policy that affirmatively advances American interests in areas where the Soviet Union is not involved.

Support for such a foreign policy would not be difficult to generate. There has been remarkably consistent support during the late 1970s and early 1980s for the American alliances with Western

Europe and Japan. In polls commissioned by the Chicago Council on Foreign Relations, it has been shown that defending the security of NATO allies continues to be of great concern to the majority of the public, that well over 70 percent consider Japan a nation vital to U.S. interests (even considering the severe economic competition with Japan), and that support for the defense of the Western Hemisphere in the Central American–Caribbean region ranges from 60 to 80 percent.[25]

At the same time, the American humiliation in Iran, the foreign policy defeat in Vietnam, the growing military strength of the Soviet Union, and intense economic competition from abroad revealed an America that entered the 1980s with a sense of control slipping from its grasp. As a response to this feeling of losing control, the public apparently wanted the United States to become more assertive in foreign policy, a feeling that played into Ronald Reagan's 1980 campaign. Ironically, as the public opinion firm of Yankelovich, Skelly, and White has pointed out,

> the present public mood would have fitted better with our position in the world of the 1950s and 1960s than it does in today's circumstances. At that time, our economic and military dominance was unchallenged —we accounted for a staggering half of the world's total industrial production. Then we could afford to adopt a more casual attitude toward the subtleties of balance-of-power politics, the will of other nations, and the limits of the American economy. But in the world of the 1980s, it may prove extraordinarily difficult to execute bold, simple initiatives that vindicate American pride.[26]

This observation has proved to be accurate, as demonstrated by the fiasco of the Lebanese intervention, the continued lack of a dramatic breakthrough on arms control, and the inability of the administration to overthrow the Sandinista government by proxy forces. The invasion of Grenada in October 1983, though hobbled by elements of military ineptitude, was a success for American arms and for certain American foreign policy objectives, but the size of the stakes involved could hardly be said to have justified such a bold assertion of American power.

Unwilling to define America's vital interests less broadly and unable to gain a consensus on the goals of U.S. foreign policy, neither the Carter nor the Reagan administration has been able to overcome the sentiment for restraint that leads members of Congress and presidential candidates to call for a facile policy of military noninterventionism, with little or no reference to the nature of national interests. The Democratic party, for example, adopted a foreign policy plank at its convention in July 1984 promising that American troops would not be sent to any region "where our objectives are not clear . . . where our objectives threaten unacceptable costs or unreasonable levels of military force . . . where the local forces supported are not working to resolve the causes of conflict."[27] This led Georgia's Senator Sam Nunn, one of the Democrats' leading authorities on defense matters, to declare: "Of course no one wants to send troops anywhere, but to declare that we won't only makes it more likely that we may have to."[28] Later, Secretary of Defense Caspar Weinberger echoed the Democrats' desire for strict prerequisites to military intervention in his own "Weinberger Doctrine," which listed a half-dozen conditions—among them, centrality to national interests and broad popular support—that must be identified before the United States takes any action abroad. What many failed to appreciate at the time was the impracticality of Weinberger's scheme: If it were applied with the scrupulous rigor he says is necessary, virtually all U.S. interventions would become impossible.

The paradox that has bedeviled American foreign policy since the failure in Vietnam is that the American people seem to want their country to be assertive, but they are unwilling to entrust the executive with the power to commit military force to gain foreign policy ends. The problem for American policymakers, and especially for an American president, is to define these goals carefully, in the light of vital national interests, thereby building public support for the deployment of American power. The indiscriminate citation of interests that is characteristic of the Reagan administration is simply not credible.

American power and interests have generally been in harmony for most of the country's history, except between this century's two world wars. Only in the interwar period, when America isolated itself from Europe, did the United States display such a singular disparity between power and interests. The United States had emerged after Versailles as a major actor in maintaining the European balance of power; it was the promise of American guarantees on which Clemenceau's postwar system of alliances depended. When America withdrew from European entanglements, the system was doomed. Then, American power outweighed its interests; now,

American interests, if they continue to be indiscriminately defined by the Reagan administration, will exceed the power of the United States to advance them.

Vietnam was always of marginal strategic interest to the United States, but it was defined as paramount. Lyndon Johnson believed not merely that the loss of Vietnam would end his presidency, but that it would "damage our democracy."[29] Recently, the Reagan administration has attempted to invest Central America with similar importance, but the broad public has not shared this assessment. Nor has the Congress, which reflects this public skepticism. And it is no longer easy for a president—even one as persuasive as Reagan—to convince the majority of the people to accept inflated definitions of the U.S. stake in every corner of the world.

To define America's vital interests more realistically and to calibrate the power to defend them is to learn to live with vulnerability. And vulnerability, as we have seen, is both the natural state of any world power and especially characteristic of American sentiment through the country's history. Too often Americans have searched for a stance that would somehow render them invulnerable.

The classic American writers understood this longing and the dangers attending it. They knew that this vision of an uncharted world where everything is possible was dangerously simplistic—that, on the contrary, everything is impure, even America, and everything is limited, even American possibilities. Nathaniel Hawthorne and Herman Melville warned Americans against being too certain about their situation or their possibilities. Melville, whose masterpiece, *Moby Dick,* was dedicated to Hawthorne, at first appears to be the quintessential American man of action. "We are the pioneers of the world," he declared in *White Jacket,* "the advance guard set on through the wilderness of untried things, to break a path in the New World, that is ours." He is attracted by the purity of an Eden, glimpsed for the first time in the South Sea islands. But later Melville's tales darken. In his story "Benito Cereno," one hears the tale of an American sea captain, Amasa Delano, who comes upon a drifting Spanish slave ship and, innocently, boards it to help. What he does not realize is that the captain, Benito Cereno, has been made captive by the slaves, who have revolted and seized the ship. When Delano himself is threatened by the slaves, he asks, bewildered, "But who would want to kill

Amasa Delano?" Unwittingly, he has been drawn into the evils of the Old World. Experience, in the guise of the Spanish sea captain, is akin to corruption; the revolt of the slaves is like a rush from darkness into light. Yet, paradoxically, that revolt threatens the enlightened American's life.

Experience should have taught Americans the dangers of innocence, but throughout the nineteenth century the U.S. government rarely admitted anything less than a moral vision of the world, in which Americans, virtuous and right, sought perfection on a continent whose vast natural resources seemed to promise autarky and, hence, invulnerability. "However our present interests may restrain us," wrote Thomas Jefferson to James Monroe in 1801, "it is impossible not to look forward to distant times when our rapid multiplication will expand itself beyond those limits, and cover the whole northern, if not the southern continent, with a people speaking the same language, governed in similar form, and by similar laws."[30] This vision inspired presidents from James Madison to Theodore Roosevelt, and by the end of World War I it seemed to many Americans that they could possess both virtue and invulnerability. When that failed, they did not retreat into isolationism—the 1920s, it should not be forgotten, found the United States intervening repeatedly in the Central American–Caribbean region—but embarked on a policy of global engagement, still in pursuit of perfectionism and invulnerability. In this respect, the country did not heed the warnings of Franklin D. Roosevelt, who counseled:

> Perfectionism, no less than isolationism or imperialism or power politics, may obstruct the paths to international peace. Let us not forget that the retreat to isolationism a quarter century ago was started not by a direct attack against international cooperation but against the alleged imperfections of the peace.[31]

America is vulnerable today and will remain so in the future. To seek strategic superiority over the Soviet Union in space will likely prove as chimerical as on earth. The economic challenge from Japan, Europe, and certain regions of the Third World is likely to increase rather than diminish. The nostalgia for those brief twenty years after World War II when American power and plenty seemed endless is dangerous. As the French sociologist Michel Crozier has put it, "Such happy days will not come back, because the dream of them is a dream about lost innocence. . . . Innocence is out of fashion for adult nations as it is for adult human beings."[32]

After the experience of Vietnam, America must act as a mature power and learn from its failure as Europeans have learned from theirs in the past.

If, as Crozier urges, the United States learns to put aside its "dreams of lost innocence and superiority," it will also find itself less estranged from a world it helped to construct, a world bound by its alliances in Europe and Asia. If America's vital interests have not fundamentally changed for almost half a century—and they have not—the means by which it is reasonable to defend those interests have altered. To define those means is the task of the makers of American foreign policy, but this can be accomplished only by an honest reckoning of America's commitments and its capabilities.

11 The American Millennium

FRANCES FITZGERALD

Frances FitzGerald is the Pulitzer Prize–winning author of *Fire in the Lake: The Vietnamese and the Americans in Vietnam* and, most recently, *America Revised*, an examination of children's textbooks used in the United States. She is a frequent contributor to the *New Yorker*, *Esquire*, and *Harper's*. Ms. FitzGerald's Vietnam book also received the National Book Award and the Bancroft Prize for history.

IF THE GHOSTS of Franklin Roosevelt and George Marshall had returned in 1984, they would have, it is certain, found the role of the United States in the world a good deal changed since their time. They would have found American power and influence remarkably diminished, in spite of the growth of a vast strategic military arsenal. They would have also found that the country had abandoned many of the policies and principles which in their view contributed to its greatness: the search for peace in a just world order, the search for a stable international economic system and for an end to poverty, ignorance, and disease. In addition, they would have found that the temper of the nation's leaders had changed quite considerably: Instead of a jaunty self-confidence or a benign paternalism, they would have encountered a defensive churlishness combined with adolescent self-promotion—a tendency to brag and strut.

As to specific foreign policies, the ghosts would have found many of them inexplicable. U.S. officials were now calling the government of a small Central American country a grave threat to American national security. They were sniping at many of the international bodies the United States had created; and they were using the United Nations as an emotional outlet for their grievances against the small, poor countries. The ghost of FDR would, of course, have chuckled at the image of the spiritual heir to Robert Taft visiting Peking and trying to sell arms to the Red Chinese; he would at the same time have wondered why the same president seemed unable to make any sort of deal with the Soviet leaders. A certain solipsism, the ghosts might note, was now abroad in the land. The American public apparently believed that the invasion of a Caribbean island 120 square miles in area was a major victory for the United States and a proof that it was now "standing tall." As for the level of the foreign policy debate, it had descended to the point where the Republican president could not explain what eighteen hundred U.S. marines were doing in Beirut when a group of unidentified fanatics blew up their headquarters, killing 241 of them, and his Democratic opponent could not present a set of policies easily distinguishable from that of the president.

Seeking some explanation for the decline of American power in the world, the ghosts might wonder whether the Soviet Union had not won the Cold War and forced the United States into retreat. Looking east, however, they would find that the Soviet Union had not gained very much ground. It had recovered from the war, of course, but its economy was still not performing well. The Soviet leaders had put a high proportion of their resources into building a military arsenal, and they had attained strategic parity with the United States. This strategic arsenal, however, brought them largely theoretical advantages; practically speaking, it served them no better than the American strategic arsenal served the United States.

The postwar military lines were still in place in Europe, and the Marshall Plan had worked even better than its creator dared to hope. In 1946 there had been some reason to believe that the Western democracies might collapse and, under pressure from their communist parties, move into the Soviet orbit. But now the political tide had turned; it was only Soviet tanks which kept most of the East European countries within that orbit. On their eastern borders the Soviets were more or less blocked. In 1950 it was said by many that the United States had "lost China"; now it appeared that the Soviet Union had lost it far more decisively. The one advance the Soviets had made was into Afghanistan, but in that harsh country small bands of ragtag guerrillas had tied down more than a hundred thousand Soviet troops for five years, and their client regime was still too weak to stand on its own. In the rest of the world the Soviets had developed a few client states, but none of any reliability except for those whose economies they were subsidizing with vast sums of money (in Cuba's case, $3 billion a year). The Soviets could not afford any more friends like Vietnam and Cuba; they had said as much to the Nicaraguans.

Finally, and most important, the life had gone out of Soviet Marxism-Leninism. The Soviet leaders still trumpeted the old phrases, and there were still a lot of excited young men around the world willing to repeat them for the price of a Kalashnikov, but the words had become little more than white noise. The communist parties in Western Europe had withered, and Soviet-style economics had proved unworkable, particularly in the nonindustrialized countries. What remained of Leninism was Lenin's analysis of revolutionary conditions and the structure of the party, as a management system more efficient than most military dictatorships. But as a language and a faith, communism had proved too thin a gruel to sustain the life of any nation. George Orwell had envisioned a world where giant bureaucracies erased historical memory; but by 1984 the new revolutionary movements sprang from religious and cultural reaction.

In fact, the world of 1984 was by and large the very oppsite of the one Orwell had sketched in 1948: instead of two or three great empires, a world of more than a hundred and fifty nation-states and dozens of separatist movements—from the Eritreans to the Basques and the Sikhs. And instead of two or three vast bureaucracies managing passive populations, an endless succession of rebellions, revolutions, coups, and civil wars, plus an extraordinary clamor of voices demanding one thing or another.

This constituted the real change since the days of Marshall and FDR. Until World War II the colonial powers had ruled half the planet without great expenditure; the British had governed all of India and the French all of Indochina with fewer than a hundred thousand troops between them. That was the world Orwell grew up in—where national politics had been the affair of small elites and the great mass of the people knew only their own tribes and villages. But then an explosion occurred; the traditional social structures were swept away, and the kings, tribal chieftains, and former colonial functionaries along with them. The new politics brought forth a strange array of new leaders—Indira Gandhi, Fidel Castro, Muammar al-Qaddafi, Lee Kuan Yew—but whether the result was good government, bad government, or no government at all, these countries were now astonishingly resistant to control from the outside. In the 1980s the Soviets could not pacify Afghanistan, the Israelis could not control southern Lebanon, and it was all that the United States could do to occupy a small island in the Caribbean. In the 1970s American statesmen made plans to defend the Persian Gulf against a possible Soviet move to cut off the so-called economic lifeline of the West. In the 1980s the United States and the Soviet Union stood on the sidelines as observers to a war between two Third World countries, Iran and Iraq, over that same gulf, neither having a firm idea of what was happening on any given day and, worse, neither knowing which country it would prefer to see win.

It was understandable that the Russians would not fare very well in this new world, for with all their talk of "progressive forces" and "the socialist brotherhood," they had a peasant conservatism and an authoritarian turn of mind which rejected both novelty and equality. Their history did not prepare them for the chaos of the United Nations any better than it prepared them for the electronic information age. The Americans, on the other hand, should have found it, if not an ideal world (it was hardly that), at least a familiar one. With its outlaws and vigilantes, it had some smell of the old American West; it was a world of market forces unrestrained, and it was, for better or worse, a fulfillment of President Woodrow Wilson's hope for national self-determination. Why, then, did Americans seem to find it so dangerous and uncomfortable a place? "We live in an age of anarchy both abroad and at home," President Nixon had said in April, 1970.

> We see mindless attacks on all the great institutions which have been created by free civilizations in the last 500 years. Even here in the United States great universities are being systematically destroyed. Small nations all over the world find themselves under attack from within and from without. . . .
>
> If, when the chips are down, the world's most powerful nation, the United States of America, acts like a pitiful, helpless giant, the forces of totalitarianism and anarchy will threaten free nations and free institutions throughout the world.[1]

This was Nixon speaking, a westerner from a town not a century old and a man who in his time attacked the freedom of a number of institutions, including the University of California at Berkeley (on whose faculty he had wished to impose a loyalty oath). What he was speaking about here was not a barbarian invasion of New England but rather his own order for a small operation by American troops across the border from Vietnam to Cambodia.

Asked to locate the major turning point for the United States since World War II, most Americans would point to the Vietnam War. The defeat was a national trauma from which the national psyche had hardly recovered by 1984; that was how television commentators tended to put it. This conception, however, might have puzzled Marshall or FDR. For the United States the Vietnam War did not represent an effort in any way comparable to World War II; the reserves were never called up. True, it had gone on for a very long time and had cost a great many lives; the effort had been unsuccessful as well as unpopular. But now that it was over, no American politician seemed able to explain why the United States had sent troops to Vietnam in the first place, or what it had lost by failing to win. "Credibility" was the usual answer. But what did that mean? The defeat had no effect on national security as defined by real assets such as sea-lanes or markets—or even allies in Southeast Asia. Indeed, arguably, the United States was now in a better position in Asia, militarily, economically, and politically, than it had ever been before.

What, then, was the explanation for the strangely erratic nature of American foreign policy? What explained the fear American politicians seemed to have of the Third World and the sense of imminent crisis that inhabited their rhetoric? Small problems seemed to inspire enormous reactions, while the serious, long-range problems got no attention at all. The fact that politicians and other commentators described the consequences of the Vietnam War in psychological, rather than political, terms suggested that the problem was subjective. And, indeed, a good many foreign policy experts thought this was the case.

Henry Kissinger, for one, confronted the question directly in his memoirs. During his term in office, he wrote, he had proposed a consistent foreign policy involving a "permanent exertion" on behalf of two long-range goals: maintaining the "global balance" and avoiding nuclear war. His grand design had, however, been thwarted—on one hand by his president, who unaccountably was ruined by the Watergate scandal, and on the other hand by the American people, who could not understand the subtlety of his approach to the U.S.-Soviet relationship and who would not commit themselves to the "permanent exertion" necessary to sustaining it. He now thought it possible that the American people might be by their nature incapable of making a "permanent exertion" on behalf of long-range foreign policy goals, since throughout their history they had been caught up in a cycle of "exuberant overextension and sulking isolationism . . . each conceived in moralistic terms." The idea, he said, was not original to him. And, indeed, he was far from alone in this belief. In foreign policy circles it was conventional wisdom that the United States had historically alternated between moralistic interventionism *à la* Woodrow Wilson and moralistic isolationism *à la* Robert Taft. It was also conventional wisdom that this was the source of the problem and the reason why the country could not establish a stable, coherent foreign policy.[2]

The historical argument was, however, mostly nonsense. From the founding of the Republic through World War I, and even, arguably, World War II, American foreign policy had been perfectly stable. The country had pursued long-range interests with persistence, and with remarkable success. In recent years historians had taken to saying that President Washington had announced an "isolationist doctrine" in his Farewell Address. He had done nothing of the sort. What he had said was that the United States should "observe good faith and justice towards all nations" and "cultivate peace and harmony with all," avoiding "permanent, inveterate antipathies against particular nations and passionate attachments for others." Europe, he had continued,

> has a set of primary interests which to us have none or a very remote relation. Hence she must be engaged in frequent controversies, the causes of which are essentially foreign to our concerns. Hence, therefore, it must be unwise in us to implicate ourselves by artificial ties in the ordinary vicissitudes of her politics or the ordinary combinations and collisions of her friendships or enmities.
>
> Our detached and distant situation invites and enables us to pursue a different course. . . . It is our true policy to steer clear of permanent alliances with any portion of the foreign world. . . .

What Washington counseled was not isolationism but rather the pursuit of American interests independent of the revolutionary and monarchical superpowers of Europe. What he counseled, as Gary Wills has pointed out, was a strategy of nonalignment, much like that adopted by Nehru, Sukarno, and other Third World leaders after decolonization. Washington had to make more of a case for it than Nehru did, because of the thinness of the cultural membrane that separated Americans from Europeans. "I want an *American* character," he wrote, "that the powers of Europe may be convinced that we act for *ourselves* and not for *others*."[3] What Washington and his immediate successors wanted specifically was freedom for American shipping in the Atlantic, European markets for American goods, and freedom of opportunity for American traders and settlers in the New World. The doctrine did not preclude temporary alliances for these ends, but it very specifically rejected the making of alliances for emotional (or moral) reasons in favor of a rational pursuit of national interests.

Washington's successors heeded this testament and pursued a consistent and coherent foreign policy for a century thereafter. There were no cycles of "withdrawal" or "overextension"; rather there was a "permanent exertion" of the sort Kissinger would have approved, in two directions. Nonalignment vis-à-vis Europe was one, and the other was expansion to the west and south. Even in

Jefferson's time the acquisition of new territories was not just a private matter; it was government policy, and one which, given the European colonies and the indigenous peoples surrounding the United States, could not possibly be called isolationist. Jefferson wanted Canada and Cuba; he got Louisiana instead. His successors, one after another, moved the borders of the nation out by means of war, treaty, and purchase. With the frontier closed, they pushed across the Caribbean, into Central America and across the Pacific.

In 1823 President Monroe added a third strand to American foreign policy. What the Monroe Doctrine meant in detail would be debated by succeeding administrations, but what it was in effect was a combination of the other two policies applied to the southern part of the American hemisphere. Its aim was quite simply to push the Europeans out of the hemisphere as far as they could be pushed, and, like the other two policies, it could in no way be called isolationist.

In fact the term *isolationist* came into currency only after World War I, and then as a term used by self-styled "internationalists" to derogate those who opposed American participation in the League of Nations and later in the war against Hitler. But it was always a misnomer, for those politicians who opposed intervention in Europe were, generally speaking, the most ardent interventionists when it came to the Caribbean, Central America, and the Far East. Republican Senator Henry Cabot Lodge of Massachusetts, for example, opposed the League of Nations treaty but rivaled Theodore Roosevelt in his enthusiasm for the Spanish-American War, the building of naval stations across the Caribbean to the Pacific, and the annexation of the Philippines. What isolationism meant was merely a tropism to the south and west rather than to the east, and it was to a great degree a regional phenomenon. Midwesterners, whether Progressives or conservative Republicans, tended to favor wars on the opposite side of the continent from easterners. While New Englanders sometimes opposed western projects, such as the war against Mexico and the annexation of the Philippines, westerners had no appreciation of American interests in Europe. In the 1930s the historians Charles and Mary Beard identified these two schools of thought more precisely as "collective internationalism" and "imperial isolationism."[4]

Conventional wisdom to the contrary, the vagaries of current foreign policy have little to do with the history of American foreign policy before World War II. They have, however, a great deal to do with what the "collective internationalists" and the "imperial isolationists" said to each other when they finally met. They have also to do with the popular religious vision of American foreign policy and the role of the nation in the world.

As the appellations imply, the difference between the two groups was more than one of geographical inclination. By the turn of the century they had very different pictures of the world—and the role of the United States in it. The collective internationalists, largely Democrats, looked east and saw a continent of heavily armed nations with large empires, strong economies, and strong cultural ties to the United States. Europe might be on the decline, but it was the civilized world. Beside it, the United States was a "developing country" and not a little provincial; until World War I it depended on European capital, on European science and technology.

The imperial isolationists, on the other hand, most of them Republicans, looked west and saw stretches of territory inhabited by peoples of alien cultures, all of them technologically and militarily inferior to the United States. The businessmen looked upon Central American and Asian countries as producers of raw materials and new markets for manufactured goods; the missionaries looked upon them as huge reservoirs of unsaved souls. In the view of the imperial isolationists, the United States was becoming the preeminent power in the world; it had no peers, it had no need for alliances, and it had almost no need for diplomacy. Pearl Harbor brought the isolationists into World War II, and after it, they could not conceive why the United States had "lost China" to Mao Zedong. Senator Robert Taft, the isolationist par excellence, supported military intervention on behalf of Chiang Kai-shek and supported MacArthur's drive into North Korea.

While the isolationists pursued one of the two "permanent exertions" of American foreign policy, the internationalists pursued the other. But much earlier than the isolationists, they faced the need for a change of tactics, even strategy, to accommodate changed circumstances. The Industrial Revolution had compressed the distance between the United States and Europe; how then should the United States pursue its long-term goals? Woodrow Wilson, generally thought of as the iconoclast of traditional foreign policy and almost the inventor of "exuberant overextension conceived in moralistic terms," acted, it could be argued, along fairly traditional lines. In August 1914, at the outset of World War I, his administration proclaimed American neutrality and demanded freedom of the seas for neutral nations. This was the United States' traditional claim, and the European nations responded as they usually did in extremis by rejecting it. England, the preeminent naval power, put this lack of agreement into practice by cutting American trade with the Central Powers. Wilson refused to try to embargo England, though his secretary of state, William Jennings Bryan, insisted that this was the only course consistent with American neutrality. Wilson's sympathies lay with the Allied Powers, it was true. But Wilson also knew, as his predecessors had since Waterloo, that when England went to war, the United States had the choice of complying with its trade restrictions or having no trade at all. Under the circumstances, it was Bryan's insistence on "neutral rights" that was quixotic and self-defeating.

Then, too, whatever Wilson felt about the democracies, he brought the United States into the war only at the eleventh hour—and then with conditions for the peace. Among his Fourteen Points, a number, such as freedom of the seas and a limitation on naval armaments, embodied traditional American foreign policy goals. Others, such as the creation of a League of Nations, could be read as attempts to adapt traditional American foreign policy to the twentieth century. Major wars in Europe (not the ordinary vicissitudes and conflicts) had always been bad for the United States, and now that military technology had made them at the same time more likely and more threatening to the United States, the attempt to create a stable order in Europe could be justified as practical, hardheaded policy.

Wilson, as it turned out, was absolutely right in predicting that a peace based on military victory and demands for large-scale reparations would be a peace founded "upon quicksand." He couched his demands for national self-determination and democracy in absolute terms, and he called himself an idealist on this score. All the same, his attempt to remake Europe at the Paris Peace Conference could be interpreted as a kind of Progressive trust-busting operation designed to break up the great power monopolies so that they could not restrain trade or start another world war. Where he was unrealistic was in his belief that the Europeans would fall in line with his plan and in his hope that the isolationists could be convinced that the United States would have to fight in the next major European war.[5]

In fact, it was only after World War I that the "permanent exertion" of the United States faltered—and then, it is well to note, on the European front alone. In the 1930s, however, the failure of the United States to respond to Hitler did not appear to be a departure from established policy—even to the internationalists who favored a response. Only after the war did the internationalists see it as a major abdication, and only then did they feel the guilt of Munich devolve upon themselves. What was now required, in their view, was a major, sustained effort to see that a "Munich" would never happen again. What they discovered, however, was that even World War II had not changed the outlook of the isolationists. When Truman and Marshall tried to persuade an important group of congressmen to appropriate $400 million in aid to Greece and Turkey, they found the congressmen loathe to spend anything for the Greek monarchy in its civil war with the Communists or for the Turkish regime in its effort to resist Soviet intimidation. Acheson then tried another tack. "Like apples in a barrel infected by one rotten one," he said, "the corruption of Greece would infect Iran and all to the east," including Egypt, Africa, and Asia Minor—and, by the by, all of Europe.[6] Truman's subsequent statement that "it must be the policy of the United States to support free peoples who are resisting attempted subjugation by armed minorities or by outside pressures" might have been an attempt to make Acheson's claim sound a bit more reasonable. But Acheson had found the lever which would move the isolationists. The congressmen voted aid to Greece and Turkey on the grounds that they were the gateway to Asia.

The Truman Doctrine augured a new era in American foreign policy. World War II had changed the world in a way that even World War I had not, and thus the United States required a new definition of its role. The doctrine did not, however, provide the definition; it merely expressed a pious intent. The internationalists, of course, had their priorities: the defense of Europe and the containment of the Soviet Union in Asia Minor. And they acted on them. The Chinese might be a "free people" under Chiang Kai-shek, but that was not, for Truman and Acheson, sufficient reason to commit American ground troops to the folly of trying to rescue China from Mao Zedong. Fearing the isolationists, however, they never enunciated their priorities. This did not save them from the wrath of Senators Robert Taft and Joseph McCarthy, but it saved their policies. There seemed nothing to be gained from precision: Vague global commitments were politically the safest.

In the end Acheson's "rotten apple" metaphor applied to his own speech better than to the Greek civil war, for that speech infected the rhetoric of makers of American foreign policy for years thereafter. In the 1950s and 1960s American congressmen could not seem to appropriate a dollar in foreign aid without proclaiming it critical, crucial, or vital; American presidents could not take the most modest of initiatives (such as the formation of CENTO in 1955 or the launching of the Alliance for Progress in 1961) without assuring the Congress it would save the world from communism. Indeed, the anti-communist rhetoric seemed to swell in inverse ratio to the importance of the initiative, reaching a crescendo when it involved attempts to sell policies of the most marginal application to the Cold War. The end result was general confusion about priorities among policymakers and a corruption of the relationship between them and the public at large. While the policymakers suspected there were some limits somewhere, they continued to commit the United States to the support of free peoples everywhere, and thus to the strategy of advancing in all directions at once. Inevitably, miscalculations occurred. In 1950 MacArthur crossed the thirty-eighth parallel and brought the Chinese into the Korean War. In 1956 the Hungarians revolted in the hope that the Eisenhower administration would act on its promise to "roll back" the Soviet lines, and were crushed by the Soviet tanks. In 1963–65 the United States committed itself to a war in Vietnam that it had no strategy for winning.

The American commitment to Vietnam was, of course, something more than a miscalculation. As it occurred over a long period of time, so the motivating forces could be discussed quite sensibly in a number of ways. But any explanation would have to take account of the atmosphere of the times. In retrospect, the notion that stopping the Vietnamese revolution at the seventeenth parallel would end all "wars of national liberation" around the world, or, alternately, would stop all of Southeast Asia and Oceania from "falling to communism" must sound a bit far-fetched. In the mid-sixties, however, a great many otherwise sensible Americans took these propositions quite seriously. For twenty years the country had gone through a series of mood swings about the state of the world: on one hand, the sense that Americans could control events everywhere if they made the effort; on the other hand, the sense that if they did not, the whole free world would crumble. According to Eisenhower, South Vietnam was the first in a series of dominoes; according to

Kennedy, it was "the keystone to the arch, the finger in the dyke."[7]

The metaphors controlled—and, at least in the public domain, prevented—the usual sorts of cost analysis. In the privacy of their access codes a few high officials made the analysis with a fair degree of accuracy. In 1964, for example, Robert McNamara predicted that the war effort would require at least a quarter of a million American troops. His estimate fell short, but at the time it would have shocked the American public, for President Johnson was running for election on a platform of keeping the Communists out of Saigon and the American troops out of the war. Such was the corruption of the political dialogue that government officials dared not say what they thought to be true lest rival politicians denounce them as defeatists, soft on communism. Later the same officials would wonder why the antiwar demonstrators called them liars and attributed the worst possible motives to them.

Of course, in the early to middle sixties even those officials who saw Vietnam as a risky proposition found it hard to believe that the United States could not in the end prevail. The British and the French had gone through their traumas of decolonization, but —excepting China—Vietnam was the first place where the United States had met the new politics of the Third World. In the 1950s American intervention of the sort undertaken on behalf of Ngo Dinh Diem had proved successful in a number of "underdeveloped" countries. In South Korea and the Philippines, for example, the United States had bolstered—even created—friendly governments and damped down insurgencies. In other parts of the world, it had dispatched unfriendly governments with little effort and little publicity. It took the CIA only three weeks to get rid of and replace the government of Jacobo Arbenz in Guatemala; it took one CIA operative only three days to depose the government of Muhammad Mossadegh and restore the Shah to the peacock throne in Iran.[8]

South Vietnam was different in that it had a strong communist movement with a secure rear area, but why should Ngo Dinh Diem not succeed there as Syngman Rhee had in Korea, creating conditions for a permanent partition of the country? Even in 1965, with Diem gone and the writ of the Republic of Vietnam running hardly to the airport, most officials believed American forces plus strategic bombing would stop the Communists. General William C. Westmoreland believed it. Even Robert McNamara could not predict with certainty the war could not be won, not with a half a million American troops, an open-ended commitment, and as much latitude as any government has in fighting a limited war.

By 1972 certain realities had hit home in official Washington, not just the reality of military power in Vietnam but other realities such as the Sino-Soviet split, the growth of the Japanese and West European economies, and the new-found leverage of the oil-producing countries. Recognition that the United States had to redefine its role came not only within the McGovern wing of the Democratic party, but also within the old foreign policy establishment and the Nixon administration. The way prepared by the National Committee on U.S.-China Relations, with encouragement from prominent businessmen, Kissinger and Nixon went to Peking. They then went to Moscow to discuss a number of new subjects. And in the wake of the 1973 Yom Kippur war, Kissinger went to the Middle East to start work on a comprehensive settlement between Israel and the Arabs. The three initiatives were groundbreaking. In statements accompanying them, Nixon and Kissinger made it clear that anti communism was not the only goal and motivating force of American policy. The United States had an interest in a relationship with China, it had an interest in certain limited forms of cooperation with the Soviet Union, and it had an interest in a just peace in the Middle East for its own sake.

What American interests were in the rest of the world remained, however, unclear. The ink hardly dry on the January 1973 Paris Agreement on Vietnam, Kissinger called on the nation in Achesonian rhetoric to advance by force in all directions at once. He urged Congress to fund covert operations in Angola, adopted the Shah of Iran as a military surrogate in the Middle East, decided to "destabilize" the government of Chile, and was just barely restrained from ordering covert operations against the first democratically elected government in Portugal. Kissinger argued that all these initiatives would have been huge successes had it not been for Congress and the new cycle of "sulking isolationism . . . conceived in moralistic terms." Before 1973 he quite understandably preferred to think of the student protests against the continuing slaughter of Indochinese as "moralistic." Less understandably now, he painted Congress's refusal to follow him into Angola as a failure of American will. Did he really believe that? Apparently so. That Congress would question his desire to intervene in Angola or stake American prestige on an attenuated oriental despotism in Iran did not strike him as rational, not even in the middle of an economic crisis brought on by deficit spending on the Vietnam War and the victory of the oil-producing nations of OPEC in a price war led by the Shah himself.[9]

If there was anything new in the Nixon-Kissinger approach to the Third World, it lay in the Nixon Doctrine, the notion that governments of large and (for the moment at least) friendly countries could be persuaded to provide the troops to defend American interests in their regions.[10] A pale copy of the Truman Doctrine, it promised that through surrogates the United States could intervene everywhere at once. At the same time it seemed, when announced, to be the *n*th term of the Achesonian series; it had the sense of an ending and the premonition of a new multilateralism.

As it turned out, however, faith in the efficacy of thin-red-line tactics continued unabated in other quarters. A few years later, when the Carter administration was reeling from the overthrow of the Shah, the Brzezinski wing of it advocated the creation of a "rapid deployment force" and prepared to do battle with the Soviets, or their surrogates, in the Persian Gulf and the Horn of Africa. For Brzezinski, as for Kissinger, the Iranian revolution was caused by American guilt over the Vietnam War. But then, instead of the Soviets, the Ayatollah Khomeini appeared, and elements of the rapid deployment force, sent to rescue the hostages, ended in a sand dune far from Tehran. Diplomacy eventually saved the hostages, but in the United States the very use of diplomacy was understood as a sign of American weakness, and though alive and at home in the United States, the diplomats remained "hostages" politically. Elected at least in part on the issue of American strength, the Reagan administration set out to follow the traditional formula for intervention. It constructed a domino theory for Central America, sent military advisers to El Salvador, and backed the "contras" against the Nicaraguan government. Secretary of State Alexander Haig then decided to bolster the Gemayel government in Beirut with eighteen hundred marines. Finally, in October 1983, Reagan found an enemy—in Grenada—small enough to overcome.

Looked at one by one and on their merits, American military initiatives in the Third World over the past twenty years have an Alice-in-Wonderland quality to them. What were the marines supposed to do in a country where every young man has a gun? Was the CIA to be charged with blowing up Gulf Oil installations in Angola? The initiatives make sense only in relation to a Tolkien vision of the world—to a mythical map. On this map there are only two countries with any power: the Soviet Union and the United States. The two are engaged in a global military, economic, and ideological struggle for control over the whole rest of the world, and such people as Ho Chi Minh, Ferdinand Marcos, and Yasir Arafat are merely pawns in the great game between the two superpowers. All of the hundred and sixty-odd nation states in the world belong to one side or another, or they serve as battlefields between the two. This map appeared in some literal detail in American junior high school textbooks of the 1950s (the communist countries painted black, the free world countries white, and the neutral nations grey). Since then the map has dropped out of the textbooks, because keeping up with the changing status of Third World countries apparently proved too Orwellian a task for the publishers. But in the minds of many Americans, including a number of those in public office, the map has remained largely intact to this day. In political debate the one question is whether it describes an ideological, or merely a geopolitical, reality.

Where this map came from and why it so appeals to Americans in spite of the cognitive dissonance involved are perhaps the key questions of American foreign policy. The first question is easily answered, for there is a well-marked trail in American intellectual history leading to the prototype. In the 1880s and 1890s those Protestant theologians who took their stand on biblical inerrancy read the books of Daniel and Revelations as a detailed set of prophecies for the events leading up to the return of Christ and the beginning of his thousand-year reign on earth. According to these prophecies, the Beast and the Anti-Christ would appear to gather the heathen nations together and assemble a huge army; there would be a period of Tribulations during which the faith of Christians would be tested, and then a world war between the heathen hordes and Christ's army of saints, culminating at the battle of Armageddon. The theologians identified Russia as the biblical land called Ros, where the Beast would appear, and America as the country that would provide the armies of the saints, since America alone contained the saving remnant of Bible-believing Christians.

Premillenialism, the doctrine undergirding this chiliastic speculation, was in a sense the most other-worldly of theological positions, but since the prophecy was for imminent crisis in the world, it became a means of interpreting current politics. In the 1920s some fundamentalist preachers saw the Bolshevik revolution as the beginning of the Tribulations; putting socialists at the head of their enemies' list (a list which then included Catholicism, along with Darwinism and the new biblical scholarship), they contributed their apocalyptic sense of cultural crisis to the "red scare." In the 1950s fundamentalists like Reverend Billy James Hargis, Carl McIntyre, and Robert Welch translated premillenialist doctrine into a description of the Cold War—and a rationale for the "witch hunts" in the United States. Then, in the late 1970s, Jerry Falwell, James Robison, and others updated the prophecies for a new crusade. "America," Falwell wrote, "is the last logical launching pad for world evangelization."[11] On a trip to Israel he walked over the prophesied battlefield for Armageddon for the benefit of his television audience. At home he called for rearmament and warned of the dangers of a communist invasion from Central America.

Premillenialism remains, of course, a minority position even among evangelical Protestants; but from time to time fundamentalist preachers have had an important influence on the American political debate. Since that influence is far out of proportion to the size of their congregations, they must be striking a responsive chord. To put it another way, elements of premillenialist thinking seem to exist in vague and diffuse form quite generally in the United States. Fundamentalist theology, for example, dictates that God and the Devil are everywhere immanent; thus politics is not simply the collision of differing self-interests but the expression of a transcendent power struggle between the forces of good and the forces of evil. Thus, if the United States is the "Christian nation," then the Soviet Union must be the "evil empire."[12]

While few politicians are actually religious fundamentalists, this kind of thinking, more or less stripped of its theology, appears quite often in foreign policy discussions. In the first place, there is the specter of global crisis evoked by American politicians at times when their European counterparts see nothing of the sort. President Nixon's speech in April 1970 on the eve of the Cambodian "incursion" is but one case in point. In that speech the Last Days were at hand and the forces of evil threatened all of civilization. However, the American people could, so Nixon insisted, halt the decline of the West by an act of will. "It is not our power," he said, "but our will and character which are being tested tonight."

Possibly Nixon was alone in seeing the Cambodian incursion as an event of global significance. But there have been other events more generally recognized as such. The rescue of thirty-nine American sailors from the *Mayaguez* in May 1975 was heralded as a demonstration that the United States had not lost its "credibility" in the world; the successful invasion of Grenada in October 1983 was seen as evidence that the United States was once again "standing tall." For these victories there would be no cost accounting—no counting the number of marines who died rescuing the sailors, no calculating the absurd disproportion of the two forces in Grenada—for the cost of signs and portents cannot be calculated. On a metaphysical battlefield any action could mean total victory; on the other hand, any faltering could bring total defeat. Thus the corruption of Greece would lead to the corruption of all Europe and Asia Minor, and the fall of South Vietnam would lead to the fall of all Southeast Asia and Oceania up to the shores of the United States. In this way of thinking there are no intermediate redoubts, no compromise positions; it is all or nothing and always the slippery slope.

This is, of course, Manichaean thinking—an ancient heresy—and the consequence is what Richard Hofstadter called the paranoid style in American politics. The logic, as Hofstadter put it, is that "unlike the rest of us, the enemy is not caught in the toils of the vast mechanism of history, himself a victim of his past, his desires, his limitations. He is a free, active, demonic agent. He wills, indeed he manufactures, the mechanism of history himself. . . ."[13] In one of the purest examples of the paranoid style, Senator Joseph McCarthy attributed the defeat of Chiang Kai-shek and the "loss of China" to communist agents within the U.S. State Department. McCarthy ended as a madman, and his charges now seem risible. But at the time a great many respectable people involved in the "China lobby" proposed a modified version of the same thing, namely, that the United States had the power to reverse the Chinese revolution and that the responsibility for that revolution lay with the Soviets, not the Chinese. Furthermore, examples of the paranoid style abound in the 1980s. In March 1981 President Reagan described American military aid to the government of El Salvador in the following fashion:

> What we're doing . . . is [to] try to halt the infiltration into the Americas by terrorists, by outside interference and those who aren't just aiming at El Salvador but, I think, are aiming at the whole of Central and possibly later South America—and, I'm sure, eventually North America. But this is what we're doing, is trying to stop this destabilizing force of terrorism and guerrilla warfare and revolution

> from being exported in here, backed by the Soviet Union and Cuba and those others that we've named.[14]

Two years later the president revealed to a joint session of Congress the probable consequences of inaction in Central America:

> If we cannot defend ourselves there, we cannot expect to prevail elsewhere. Our credibility would collapse, our alliances would crumble, and the safety of our homeland would be put in jeopardy.[15]

Of course, some secular thinking has gone into the popular debate on American foreign policy, but until recently the main alternative to the Manichaean world view has been another mode of millenial thinking that one might call Gnostic (for its brother heresy). This mode, too, can be traced to nineteenth-century Protestant theology, but in this case to the more optimistic doctrines of Transcendentalism, Perfectionism, and the Social Gospel. Theologians of the Social Gospel, for example, believed that evil was a temporary phenomenon, and that human nature was not merely perfectible but would actually soon be perfected as a result of the swiftness of moral evolution in the United States.[16] The Social Gospel was as much a minority position as premillenialism among Protestants, but the general strain of thinking was quite widespread. Woodrow Wilson went to the Paris Peace Conference with the Gnostic vision that the United States could bring peace, freedom, and justice to the world by an act of will. During and after World War II it was this vision which filled the rhetorical sails of a great many public figures from Roosevelt to Wendell Willkie to Henry Luce. Since then this mode of thinking has been well represented in American history textbooks for young students and in political speeches—mostly, but not exclusively, by liberals. (Hubert Humphrey is the best example of the liberal Gnostic.)

The Gnostic vision is much more confident and generous than the Manichaean world view. But it contains what Reinhold Niebuhr pointed to as the dubious (and un-Christian) proposition that Americans act without taint of self-interest or self-love. The notion that Americans act disinterestedly, whereas others act only out of self-interest (and therefore have no legitimate point of view), is, of course, American exceptionalism and the American imperial vision. Used to justify territorial acquisitions in the nineteenth century, it has on occasion cropped up as justification for American military undertakings in the twentieth.

In 1965 President Johnson pledged not only to prevent another

Munich in Southeast Asia but also to bring peace, material progress, and human dignity to the Vietnamese. "The American people," he said, "have helped generously in times past in these works. Now there must be a much more massive effort to improve the life of man in that conflict-torn corner of our world."[17] These promises proved extremely troublesome. They lay Johnson open to the charge of hypocrisy, and their logic permitted elements within the peace movement to conclude that if the United States did not represent absolute good, it must represent absolute evil.

Gnostic rhetoric has also been used by politicians in support of a great many peaceful and generally positive undertakings in American foreign policy, notably, in the early days, in support of the United Nations. Here, too, it has tended to backfire—though, of course, in the opposite direction. In the case of the United Nations, it turned Americans against the organization as soon as members from the Third World began to disagree with the United States, the logic being that if those countries were not America's loyal friends, they must be enemies. In the early seventies Daniel Patrick Moynihan, then Nixon's ambassador to the United Nations, made several speeches deploring the decline of Western civilization before the advance of totalitarianism and barbarism in the Third World. The United Nations, he said, consisted almost exclusively of communist regimes and "ancient and modern despotisms," all of which were united in their "conviction that their success ultimately depends on our failure."[18] The speeches were widely acclaimed and, at around the same time, large numbers of Americans confided in pollsters that they had "lost interest" in the rest of the world.[19]

Gnostic thinking has also, and most unfortunately, infected the debate over nuclear weapons, leading elements in the peace movement to conclude that deterrence is unnecessary and that the only obstacle to world peace is insanity in Washington. Kissinger was surely right when he said that it was difficult to find a constituency for the proposition that "we had to collaborate with our adversary while resisting him."[20] The Manichaeans would not accept it, but no more would the Gnostics.

Given American intellectual history, it is in a sense not at all surprising that Americans should think about foreign policy in religious terms. Every major social movement in American history, including feminism and civil rights, came initially out of the churches, and the major political changes in the country were preceded by the three Great Awakenings. Religious revivals are what the United States seems to have instead of revolutions; it is a land of prophets, not of philosophers. Clergymen of all denominations inject themselves into the political debate in a way that would be inconceivable in Europe, and Americans continue to convince the pollsters that theirs is the most God-fearing society in the Western world.

But there are various kinds of religious thinking. When groups like the National Conference of Catholic Bishops or the National Council of Churches or the Southern Baptist Convention enter the debate directly, they offer moral reasoning, and sometimes of a high order. But the kind of thinking that has governed the foreign policy debate has not been moral so much as eschatalogical, and its theology is the nineteenth-century brand. That millenial thinking should exist within one of the most technologically advanced of nations is perhaps not remarkable. What is remarkable is that it exists in a society which has proved extraordinarily good at governing itself by compromise and consensus, and whose domestic politics are ordinarily nonideological and consist instead of appeasing, co-opting, and otherwise dealing with a vast number of self-acknowledged interest groups.

There is a historical explanation for the persistence of this millenial thinking; there is also a psychological one. For most of the nineteenth century Europe and Asia were so distant from the United States as to belong to the realm of the imagination for most Americans. At that distance Americans could conceive of the nation itself as the New World, a model for humankind, the City on the Hill. After the War of 1812 with Britain, the policy of nonalignment was entirely successful; not only did it keep the United States out of European wars, but it permitted westward expansion to proceed apace. Americans could thus continue to believe that the United States was different from the European nations in its international behavior, and would eternally remain so.

Because to the south and west there were only Indian tribes and colonies (or former colonies) of Catholic Spain, Protestants saw national expansion as identical with the missionary enterprise. At the end of the century President McKinley told a group of his Methodist brethren that he favored annexation of the Philippines because it was the American duty to "uplift and civilize and Christianize" the Filipinos.[21] This evangelical mission was not just the substance of American policy; for many it was the spirit of the nation itself. Who could dispute that the nation had a God-given mission, a manifest destiny? The proof was its extraordinary success. No military defeat ever forced Americans to reconsider, or to disentangle their religious visions from their view of the nation in the world. The great and sobering disaster for the nation was domestic—the Civil War. After it, an agreement to disagree came to seem vital to national survival. In 1876 few voters complained of the cynical political deal that ended Reconstruction; it seemed a necessity, a fact of life. Abroad, however, such necessities never arose, and there Americans could continue to believe that their nation had a higher calling and the grace to do God's will.

Neither World War I nor World War II gave Americans much reason to question the exceptional nature of the United States. On the contrary, both seemed to confirm it, and the latter in particular. Hitler, after all, was as demonic as any figure in history, and fascism (as was discovered later) had loosed an almost metaphysical evil in its true believers. Stalinism seemed its successor. But while the war laid waste to every other industrialized country, it released huge energies in the United States, both human and technological. The atomic bomb and American prosperity seemed to lift the country out of the realm of necessity. Its power seemed limitless. *Present at the Creation* was the title Dean Acheson gave his memoirs of the postwar period. He was the son of a clergyman; so was John Foster Dulles.

As secretary of state, Henry Kissinger never quite got around to educating the American public "in the complexity of the world we would have to manage" and persuading it to abandon "the simpler verities of an earlier age."[22] Jimmy Carter, however, put a good deal of effort into describing the complexities of the world as he saw it during his term in office. The world, he said in effect, is a very complicated place. It is very small, very crowded, and nations have become interdependent; what Americans do affects others, and yet Americans cannot always get what they want by unilateral action. This was no Gnostic vision, nor was it a bipolar "global balance" vision of the sort which accorded so well with the Manichaean world view.

Carter's explanations, however, combined with a demonstration that he was more or less right—the hostage crisis in Iran—went over very badly with the American electorate. The traditional map of the world, as presented by Ronald Reagan, proved much more attractive. It had the virtue of simplicity. While it held certain terrors, such as subversion in the hemisphere, it was reassuringly familiar and at least not so claustrophobic. In his world Americans could afford to pay no attention to gloomy matters like overpopulation, nuclear proliferation, and the spread of deserts. They could forget whole continents for years at a time. On the other hand, if foreign governments were abusive or ungrateful, Americans could be abusive back without guilt or fear of consequences. Most important, perhaps, there was an aesthetic to Reagan's world. Instead of a chaotic mess, there was a story line, a single sweeping arch of narrative that bound all events together and made sense of them. Even if the story was not true in every particular, even if it left out the subplots appearing daily in the newspapers, it was surely better than no story at all.

In the 1984 election Walter Mondale did not dispute this. Unable to summon up the Gnostic vision—for where were the victories?—he told a watered-down version of the same story. The polls showed the public supporting him over Reagan on specific issues of foreign affairs. Perhaps it is that Americans respond to pollsters' questions with the "right" answers as they would to intelligence tests, and then they act as they please. The question left for the next election is what any candidate can say about foreign affairs that will get him elected and be true at the same time. The question, bizarre as it is, may be as profound as any in American political history.

12 Paths to Reconciliation

The United States in the International System of the Late 1980s

RICHARD H. ULLMAN

Richard H. Ullman is professor of international affairs at Princeton University. A staff member at the National Security Council and in the Office of the Secretary of Defense during the Johnson administration, he has in recent years served as director of studies and director of the 1980s Project at the Council on Foreign Relations, as a member of the editorial board of the *New York Times*, and as editor of *Foreign Policy* magazine. Mr. Ullman is the author of a three-volume history of Anglo-Soviet relations during the period 1917–21, and of many articles on issues of contemporary international relations.

AFTER ESTRANGEMENT, what? Divorce in some instances, reconciliation in others. And in still others—perhaps in most—no change: a continuation of the lack of communication and the separate paths connoted by the word *estrangement*. Are these the options for American foreign policy?

Estrangement is a concept entirely familiar from the realm of interpersonal relationships. Extended to relations among states, it is both suggestive and misleading. What it suggests, of course, is that relations among states, like relations among persons, are shaped (if not governed) by qualities such as trust, loyalty, and obligation—or suspicion, contempt, and hatred. These qualities inevitably come into play when heads of state meet. But they are also the product of the interactions of mass cultures, and of the tendency (surely universal) of the members of one human group to personalize its relationships with other groups.

In the conventional language of statecraft, however, notions like estrangement have no place. The dominant images among academic theorists of international relations remain those of opaque balls on a billiard table and of weights on a scale; the balls collide, or do not, according to the game-theoretical logic of strategic interaction, while the weights add up to a balance of power. Lord Palmerston, a leading figure in British statecraft for much of the nineteenth century, once declared that his nation had no permanent friends, only permanent interests. His formulation is often, widely, and approvingly repeated.

Like images that depict a hard shell around the state, which itself obeys laws of necessity (or "national interest"), formulations like Palmerston's convey a partial picture of reality. It is indeed the case that, despite vast improvements in communications and the ease of travel, states remain more or less hard-shelled. The shell of sovereignty makes the state different from other collectivities—such as the family, the tribe, the city—to which people address their loyalties. The arbitrary territorial divisions imposed by juridical borders still effectively compartmentalize the human relationships that rub up against them.

There are elements of paradox here. At one extreme, even states that as a matter of high policy shield themselves from outside influences of every sort—the Soviet Union is the archetypical example—find their borders penetrated in myriad ways every day. Yet, at the other extreme, even in regions as integrated as Western Europe, where states encourage an almost unimpeded cross-border flow of goods and persons and ideas, most people still have only the most superficial contacts with citizens of other sovereignties. Their impressions of other nations are likely to be formed by mass media rather than through direct interaction, and the mass media themselves—even in societies that pride themselves on allowing genuine freedom of expression—are to a large measure (a much greater measure than the media themselves acknowledge) creatures of their state and of the politics dominant within it.

For Americans, that is especially the case. The roots of the estrangement that, as the other essays in this volume suggest, characterizes so many aspects of this country's international relationships, lie in a shallow soil of misperceptions and stereotypical views of the world beyond the borders of the U.S. nation-continent. Most of those misperceptions have to do in one way or another with threat. For nearly all of the forty years since World War II, Americans have felt threatened, often mortally. They have, of course, perceived the primary source of danger to be the Soviet Union. That was the case even when the U.S.S.R. was by any objective standard far weaker than the United States. But the Soviet Union has been far from the only perceived source of threat. For years the People's Republic of China, despite its poverty and its inability effectively to project military power much beyond its own borders, seemed even more threatening, because of its proclaimed desire to turn the entire Third World against the West. And lately, American fears of revolutionary regimes in Third World countries have focused not only on Cuba, Nicaragua, and Vietnam (regarded as satellites of the Soviet Union), but also on Iran and Libya—and, on occasion, even on obviously weak micro-states like Grenada.

Yet the overarching focus of concern has been Moscow. Only the Soviet Union, after all, has possessed the ability to destroy a hitherto invulnerable American society. Virtually since the end of World War II, U.S. administrations have found themselves almost continuously preoccupied with preparing for the possibility of war with the U.S.S.R. From the vantage point of the mid-1980s those preparations—and all they have entailed for the mind sets of policymakers and the public—loom as the predominant feature of America's international relationships. Very few aspects of those relationships have not been touched (even contaminated) by Washington's preoccupation with Moscow, just as Thucydides tells us that preoccupation with Sparta shaped all of Athens' dealings with the other city-states of the Peloponnesus.

Nearly a decade ago, the Carter administration came into office hoping to wall off U.S.-Soviet relations so that Americans could throw aside the blinders of the Cold War and turn their attention to other foreign policy issues and problems. Carter and his advisers did not doubt that the rivalry with Moscow posed formidable difficulties. Rather, they tended to assume that a Soviet Union led by an aging and conservative leadership had become increasingly irrelevant to the task of coping with the other major problems on America's international agenda—pacifying the Middle East, fostering economic development in both the First and the Third worlds, promoting higher global standards for human rights, and achieving racial justice in southern Africa—and that those problems could be attacked without regard to the separate U.S.-Soviet agenda. The president and his associates soon found that they had been mistaken. By the time they left office they had provided a graphic demonstration that the relationship with Moscow colors all others, and that American policymakers must manage it before they can achieve their other foreign policy objectives.

The connection between relations with Moscow and everything else has been amplified and exaggerated by the adversarial nature of American domestic politics and by the ways in which those politics are channeled by the structure of American governmental institutions. That structure—separate executive and legislative branches and separate electoral districts (including a territorially based Senate), with fixed and only partially coterminous terms of office—minimizes party discipline and maximizes opportunities for demagogic allegations of presidential weakness in the face of threats to national security. As the historical record shows, a president's political enemies constantly make use of those opportunities.

Unhappily, however, so do his friends. The Carter administration found that its efforts to transfer to Panama control over the isthmian canal, to secure independence for Zimbabwe and Namibia, and to engineer the departure from Nicaragua of Anastasio Somoza Debayle (to cite only a few examples) were attacked by electorally beleaguered Democrats as well as by Republicans for allegedly opening doors for Soviet opportunism. In a particularly flagrant instance, the late Democratic Senator Frank Church of Idaho, then chairman of the Senate Foreign Relations Committee, attempted to counterbalance his support for the Panama Canal Treaties and to demonstrate his vigilance and toughness to Idaho voters by announcing in August 1979 the alleged discovery of a "Soviet combat brigade" in Cuba and insisting that the administration take action to expel it. In so doing he failed to assure either the removal of the brigade or his own reelection the following year. But he seriously weakened a president of his own party and contributed—perhaps decisively—to the Senate's failure to ratify the SALT II treaty with Moscow, an arms control agreement about which he himself cared deeply.

Billiard balls or black boxes cannot be estranged from one another: Estrangement therefore implies the dissolution of relationships held together by considerations that go beyond narrow self-interest or calculations regarding the balance of power. The international relations of the twentieth century, more than those of any other era, have been characterized by declarations of a solidarity that extends beyond interest. The Socialist bloc, the "Group of 77," the Alliance for Progress, the Non-Aligned Movement, NATO, the Arab League, the European Community, the "Front Line" states of southern Africa—these and many other associations are lauded by their members as communities of ideology or of shared experience rather than mere alignments of interest. Reality, however, is more harsh. On nearly every important issue these communities have been sundered by defections—in the language of this volume, estrangements.

Recent years have seen three large estrangements that have cut across the fabric of international relations. First and most profound is a growing indifference on the part of rich countries—particularly their governments—to the plight of the poor, whose problems seem overwhelming. Beset by economic problems themselves, the rich have rationalized their diminishing levels of assistance to the poor by asserting that in many instances they can do little more than supply palliatives, and that the combination of rapidly growing population, ethnic warfare, and low production that has afflicted the least developed states places many of them beyond real help.

The second estrangement is within the ranks of the rich themselves—the industrialized democracies of Europe, North America, and Japan. The tasks of economic management in complex modern societies seem so daunting that governments often give little more than lip service to the goal of international cooperation to preserve the benefits of open access. For national politicians it is the domestic context that counts; they keep their power by responding to domestic demands, not by international demand management. Therefore, despite commitments to openness and cooperation, measures to restrict flows or otherwise preserve short-run advantages have become commonplace. This is true not only along broad regional lines—the Americans vs. the Europeans vs. the Japanese —but even within the narrower arena of the European Community, whose members in recent years have increasingly resorted to "temporary" restrictive measures against one another.

Perhaps the most significant estrangement is the third, not a rift between states but the estrangement of many governments from the entire idea of cooperation through a formal structure of international organizations. The United States, especially, has outspokenly expressed its dissatisfaction with multilateral forums. Indeed, under the Reagan administration, dissatisfaction has verged on contempt. UNESCO, the World Court, the World Health Organization, the U.N. Conference on the Law of the Sea, the U.N. General Assembly itself, even ad hoc bodies such as the so-called Contadora Group (Colombia, Mexico, Panama, and Venezuela) that have been attempting to negotiate an end to hostilities in Central America—all have experienced the administration's scorn. But other governments, in Western Europe and elsewhere, have not been much more supportive. For more than two decades, the Soviet Union and its allies have refused to contribute financially to United Nations peacekeeping activities.

It is undeniable that, for the three large functions for which many persons in many countries have looked to international institutions—making and keeping peace, alleviating the plight of the poor nations, and removing barriers that might prevent multilateral economic progress—the global structure established in the aftermath of World War II has often proved seriously inadequate, if not (as in the case of the first of these functions) entirely toothless. Disaffection from international organization has grown not merely because global institutions have proved weak, nor because regional institutions have failed to be substantially stronger, but because the great powers have seldom seen it as in their interest to help them be strong. Thus, it is not surprising that for every decision that governments face, whether it is to allow foreign fleets to fish off their shores or to acquire nuclear weapons, there has been an increasing tendency to opt for unilateral rather than multilateral remedies.

It is against this pattern of more widespread disaffection that the current situation of the United States should be assessed. The preceding chapters in this volume have made the point that the United States is aloof from much of the rest of the world, and that indeed is true. That condition has given rise to much dismay abroad, and to concern among Americans who worry about the welfare of communities beyond their own borders. Yet one fact should be squarely faced: For the United States the costs of estrangement are tolerable. For Americans, if not for others, the consequences of estrangement are serious, but neither urgent nor potentially catastrophic.

That is because, with a few exceptions, action or inaction by the United States affects the rest of the international system to a much greater extent than the activities of other states affect the United States. The exceptions are important, and would include direct military (but not economic) action against the United States by the Soviet Union and drastic steps to cut off vital raw materials in those rare instances where a few states control major parts of the world supply—a shut-in of oil by Saudi Arabia would be the most obvious example. Neither exception is likely to occur, however; the potential penalties are too high. Mostly, the United States enjoys substantial immunity. Few, if any, American lives are threatened by massive arms purchases by Third World governments, protracted civil wars in Central America, or another round of cross-border wars in the Middle East. Yet all these situations have been exacerbated by the actions or inactions of recent American administrations. Similarly, no Americans will go hungry as a consequence of crop failures in the Sahel or the inability of Bangladesh's economy to grow faster than its population. And—perhaps most important of all—it is very unlikely that most Americans will directly feel the consequences of an intensified Cold War and an unabated arms race between Washington and Moscow. But Poles and Nicaraguans, Filipinos and South Koreans will all experience a diminution in their ability to pursue their private purposes as their governments respond to heightened international tension by—among other measures—imposing additional restrictions on the liberties of their citizens.

This is not to say that Americans do not experience the consequences of their government's foreign and military policies. Poor Americans certainly feel directly the reduction of government domestic programs cut in order to fund massive increases in military spending. Americans who are neither poor nor dependent on government welfare programs experience in myriad indirect ways the widening gap between themselves and the (largely black) poor created by these failures to address pressing social needs. These include heightened insecurity in the central cities and a demoralizing environment of urban blight. And the entire population will eventually feel the consequences of government budgets that funnel resources to the military sector instead of to the civilian scientific, technological, and other research and educational programs that are necessary to maintain the competitive position of the United States in the international economy.

It is more difficult to demonstrate that Americans may experience in any meaningful way the growing international gap between rich and poor. That millions of persons in a score or more Third World countries will die each year from starvation or from ethnic violence is unsettling; so is the notion that their countries may become virtually off limits for safe travel by more affluent outsiders, much like the bleakest urban ghettos of the United States. But the direct effect of these circumstances on the lives of Americans or other outsiders is surely minimal. Unlike the poor at home, the poor abroad are safely out of sight. The rich are affected by the foreign poor for the most part only when the poor gain entry—legally if possible, illegally if necessary—into the rich countries. There are, it is true, scenarios positing cataclysmic wars between an affluent North and a desperate, perhaps nuclear-armed South,[1] but these scenarios are scarcely likely to occur: The military-technological advantage enjoyed by the North will remain too great.

Meanwhile, during the decade or so immediately ahead, if the costs of estrangement are bearable for the United States, they will be bearable as well for the other advanced industrialized states and for the Soviet Union—and probably also for all but the most fragile developing states. All will do considerably less well than they might in meeting acute societal needs, but there is little likelihood that their survival as states would be threatened. That is because the international system is so inherently conservative. Despite profound changes in technology that have vastly altered many aspects of life, the nation-state remains the largest effective focus of individual loyalties. Moreover, the present and foreseeable environment contains no serious threat to the sovereign state. Ironically, the existence of nuclear weapons—the product of human intelligence and industry that could most readily bring about the destruction of the nation-state—serves as one of its most important props. By making war between the two leading powers (and also their allies) too costly to allow for risk taking, nuclear weapons have brought a stability approaching rigidity to the state system.

Underlying this argument is an assumption that should be made explicit: Despite changes in both technology and doctrine that have made nuclear weapons seem more usable by the superpowers, the probability of nuclear war between them is exceedingly small. The mutual deterrence that has governed their relationship since the 1950s will continue to do so. Neither power will permit the technological balance to shift to such an extent that the other can threaten it with the reasonable prospect of a successful, disabling nuclear first strike. That does not mean that the ongoing nuclear arms race between them is in any sense benign or, indeed, not profoundly harmful. But the harm is psychological and political rather than physical. The knowledge that each is deploying thousands of new, more accurate nuclear weapons cannot help but exacerbate other tensions and make more difficult the task of reaching an accommodation between them. Moreover, the ambitious nuclear weapons programs that both superpowers have pursued in recent years have been but a part—albeit the most lethal part—of a more general military buildup. It is reasonable to suppose that on the Soviet side, especially, but also on the American, the very large military buildup has emphasized the role of military considerations in foreign policy decisions and increased the likelihood that leaderships will reach for military solutions to political problems.

To citizens of the United States and the Soviet Union, and to Europeans and Japanese as well, it is undoubtedly consoling that those solutions are likely to be applied in the Third World, in regions like the Middle East, Central America, South Asia, and Africa. There, rather than Europe or even the Korean peninsula, are the likely scenes of superpower proxy wars and perhaps direct interventions by superpower military forces. In Europe and Korea, where the lines are clearly drawn between opposing forces and the risks of miscalculation are so high, both Moscow and Washington are likely to apply greater caution.

Even Japan, despite the often-expressed nervousness of its leaders regarding their country's vulnerability to shortages of raw materials (especially oil), is likely to suffer no more real harm from potential conflicts in the Third World than Western Europe. This is not to argue that Japan would be immune, only that if a commodity as crucially important as oil is embargoed for any length of time, all the industrialized states (including the United States) would be affected and they would be likely to take fairly drastic measures in response, including the use of military force. For that reason, as argued earlier, a protracted cutoff of oil is unlikely.

Those who will most acutely feel the dislocation of war or of a militarized society, therefore, are the citizens of the Third World states that are the likely arenas of superpower rivalry. Ironically, their predicament is made worse by the fact that their leaders frequently have no interest in altering that circumstance. In many Third World countries, those who govern have grown accustomed to depending on Washington or Moscow for the wherewithal—from infusions of cash to infusions of military trainers and even, on occasion, of combat forces—which enables them to cling to power against external and (more frequently) internal opposition. Undoubtedly, if the superpowers were to take a more hands-off posture, some governments would fall.

Between nations as between persons, estrangement grows out of disappointment, and disappointment stems from the failure of one or both parties to a relationship to live up to expectations. Over the four decades since World War II the United States has done much to arouse expectations. Those expectations have had to do with material considerations, such as security assistance and economic benefits, but they have also had to do with values, such as human rights. For years, governments and publics in nearly every part of the globe looked to the United States to organize security in their region with its own military forces, and also to supply weapons and military training. Similarly, because of its huge market for their raw materials and finished products and its seemingly inexhaustible supply of investment capital, the United States was seen as the engine of worldwide economic growth. Until the late 1960s, when the social and economic toll of the war in Vietnam became apparent, the United States was scarcely a reluctant partner in these relationships. Successive administrations in Washington shared a world view that emphasized the need to build up "situations of strength" in order to contain the Soviet Union, even if doing so might someday entail severe economic costs to the United States. That was the motivation behind American postwar efforts to reconstruct the economies of Western Europe and Japan and, later, to support the formation of the European Economic Community and the economic development of potential Third World competitors such as South Korea, Taiwan, and Iran.

Human rights were more complicated. It was apparent soon after the beginning of the Cold War in the late 1940s that while the United States would consistently espouse the liberal values that characterized its domestic society, it would support regimes abroad that were anything but liberal. Yet there were periods—most notably during the Carter administration, although it did not apply its policies uniformly—when Washington would wield its carrots and sticks so as actively to promote human rights in other societies, sometimes even at the evident expense of other American interests. Thus in this domain, too, expectations were created. However, for the most part they were the expectations not of governments or even of mass publics, but of democratic opposition movements and political prisoners and their families.

There have been instances when these two sets of expectations have contradicted one another. Typically, they have involved apparent commitments to authoritarian regimes in countries deemed to be of geopolitical importance to Washington. In Iran, for example, the Carter administration could not possibly have met both the Shah's expectations that the United States would back him against his internal opponents and the expectations of those opponents that it would apply to the Shah some of the same sanctions (suspending military assistance, curtailing arms sales) it was bringing against other regimes, mostly in Latin America, that were guilty of human rights violations. The regime of President Ferdinand Marcos in the Philippines may soon present similar contradictions, even with a less idealistic administration in power in Washington. In other instances, U.S. foreign policy objectives have been in conflict. Several administrations have used the promise of arms deliveries to win the support of "moderate" Arab governments, only to find themselves transferring even more arms to Israel as insurance against possible threats from those same governments.

It would be vastly oversimplifying to say that the United States has not fulfilled the expectations of other governments and peoples. Clearly, the rhetoric of the American internal debate has sometimes led foreign leaders to believe (or to contend that they believed) that Washington would stand by them through thick and thin. Nguyen Van Thieu in Vietnam, the Shah in Iran, and Anastasio Somoza Debayle in Nicaragua are obvious examples, and there are others for whom the moment of truth has not yet come—Marcos in the Philippines, the succession of repressive governments of South Korea, perhaps Hosni Mubarak in Egypt. Yet it is arguable that within the realm of security assistance the United States has amply met its promises both to allies and to states to which Washington has been tied less formally. It has owed leaders like these not salvation but the means by which, through statesmanship, they might assure their own survival. Moreover, expectations are not static. They adjust with circumstances, and the circumstances in which the United States has found itself over the last several decades have changed drastically. And despite changed circumstances, the United States is still at the center of the hopes of many governments and peoples. What can be done to fulfill them and, by doing so, reverse the process of estrangement?

The most important thing that any administration in Washington can do is also the most difficult and the least likely of achievement—indeed, for the foreseeable future, the least likely of even being attempted. That is to reverse forty years of thinking about the nature of the international system that has seen the United States as acutely threatened by the Soviet Union. Previous essays in this volume have made clear how an exaggerated perception of the Soviet threat has run like a red thread through post-1945 American foreign policy. That perception has had very little relationship to the actual military balance. Indeed, as we now know, the period when the balance was most tilted in favor of the United States—the late 1940s and the 1950s—was when the threat was generally perceived as most severe.

Americans—even members of the foreign policy "establishment"—tend to see their nation's foreign policy as nonideological and pragmatic, marked by a willingness to compromise and to meet other parties to any negotiation more than halfway. But it is striking how much the record of the postwar period belies that self-image, and how early and consistently American administrations pursued a policy toward the Soviet Union they labeled "negotiation from strength." For the decade and a half preceding the Kennedy administration, but also during much of the following period, that phrase meant no negotiation at all. American strength was not judged sufficient to risk compromise.

The lesson of Munich—never make concessions that a potential enemy might interpret as weakness—had been learned too well. What political scientists have come to call the realist paradigm—states, invariably insecure, seek security through power over others—has dominated thinking about international relations among U.S. politicians, officials, journalists, and academics. Politicians, such as Henry Wallace in 1948, George McGovern in 1972, Andrew Young in the late 1970s, and Jesse Jackson in 1984, who have tried to put forward conceptions of an American world role emphasizing drastically lower levels of armament and cooperation with perceived adversaries have been dismissed as ignorant or naive idealists. Wallace was called a communist dupe. In 1984 Jackson was the only candidate for a major party's presidential nomination who did not emphasize an international posture of "toughness" and "strength" as a virtue beyond question and who did not advocate enlarging still further an already swollen U.S. defense budget.

The realist paradigm is scarcely unique to the United States. Indeed, it has been the normal mode of thought, especially in countries plausibly capable of defending themselves against external attack. (Weak states, not strong ones, seem to provide an environment conducive to thinking about cooperative solutions to the problem of security!) Moreover, Marxism-Leninism is a realist doctrine par excellence. Since the victory of bolshevism in Russia in 1917, Soviet leaders and citizens have been deeply imbued with images of an encircling ring of capitalist states and with the assumption that only the growing strength of the "Socialist Motherland" keeps its enemies at bay. These images are of course mirrored by those, prevalent in the West, of Moscow as the center of a spreading world communist revolution that can be countered only by a policy of containment based on military power. Each system has interpreted the behavior of the other as providing plenty of evidence to confirm its worst fears.

Would the Soviet Union have responded in kind to American policies of appeasement? (The word *appeasement* is used advisedly, in its literal sense of pacification or conciliation; it deserves rescue from its more usual, politically pejorative connotation. And I would emphasize that I have in mind policies—planned approaches, systematically, coherently, and unambiguously pursued over time—not mere gestures.) Obviously it is impossible to know. But precisely because the United States was for so long so clearly the strongest member of the international system, it would have run fewer risks than any other member in pursuing policies that might have broken the pattern of East-West confrontation and spiraling arms acquisition. Alas, during the period when American preponderance was most marked, such policies were never tried. With respect to the two most critical issues, nuclear weapons and the future of Germany, the Truman and Eisenhower administrations refused seriously to contemplate cooperative solutions, because they seemed to involve greater risks than the pursuit of unilateral policies. Policymakers congenitally overdiscounted long-term risks. The dangers inherent in a world in which, eventually, both the United States and the U.S.S.R. would have large nuclear arsenals, or one in which two alliances would face each other in the center of Europe with the greatest concentrations of military force ever seen, seemed less—because less immediate—than the dangers of cooperating with the Soviets in attempting to ban nuclear weapons (as might barely have been possible in 1946) or to unify Germany under conditions that would have left it demilitarized, neutralized, and (because of the larger population in the western zones) noncommunist.[2]

Over the last two decades there have indeed been attempts to negotiate with Moscow, in particular over nuclear weapons. But the agreements reached have all done little more than codify some aspects of the existing balance; they have not seriously constrained either government from pursuing the nuclear programs it wished. Although both sides must share responsibility for this record, it can be argued that because the United States has consistently played the initiatory role in nuclear weapons development, different American decisions at crucial junctures could fundamentally have altered—vastly for the better—the U.S.-Soviet nuclear relationship.

Thus, in 1963 the two sides were close to an agreement banning all nuclear explosion tests. The effort collapsed when the Kennedy administration decided against fighting the military nuclear establishment (which was, of course, anxious to continue testing new warhead designs) over the adequacy of Moscow's offer of restricted on-site inspection, and settled instead for a ban on explosions in the atmosphere but not underground, thus enabling the development of new generations of weapons.[3] Here was a case of the United States rejecting negotiated constraints. Subsequently it went ahead and rushed the flight testing of new weapons—in 1968 multiple, independently targeted reentry vehicles (MIRVs); in the early

1980s ground- and sea-launched cruise missiles; and in late 1984 antisatellite weapons (ASAT)—so that negotiations to forbid their deployment by prohibiting their testing would be impossible. (These were all weapons whose presence or absence cannot reliably be determined except through highly intrusive on-site inspections that neither superpower would be likely to tolerate. Therefore, since flight tests can be detected, the only way to be certain they have not been deployed is if they have not been observed being tested in flight.)

Unilateralist tactics such as these may secure Washington a temporary advantage until the Soviets catch up. But American technological advantages (as the atomic bomb itself demonstrated) have invariably proved short-lived. And repeated attempts to realize those advantages have not only greatly complicated the task of enhancing strategic stability through arms control but, more generally, they have emphasized conflict in the U.S.-Soviet relationship at the expense of cooperation.

Specialists who earn their living worrying about that relationship have long debated the degree to which arms control negotiations have been—and should be—linked to other issues of East-West contention. Because arms control is most of all about reducing the risk of nuclear war, it has acquired a symbolic importance disproportionate to its results. When arms control negotiations are making progress, it is often argued, progress is easier on seemingly unrelated issues; when they are going poorly or not taking place at all, the two political leaderships find it difficult to reach agreements of any kind. Conversely, during periods of U.S.-Soviet tension over contentious issues such as intervention in the Third World—the Soviet invasion of Afghanistan is a prime example—ongoing efforts at arms control bog down.

Linkage or none, the record of success in U.S.-Soviet negotiations outside the sphere of arms control (indeed, many would argue, within it as well) is not impressive. Most evidently absent is any agreement other than at the level of rhetoric over the avoidance of superpower crises arising from political involvements in the Third World. At their May 1972 summit meeting, President Nixon and Chairman Brezhnev put their signatures to a statement detailing the "Basic Principles of Relations" between their two governments. Its purpose, as Nixon subsequently explained, was to provide a "road map" of the path they should follow to preserve peace. But its precepts were so general as to provide no bulwark against the crisis that occurred when the two superpowers nearly collided during the October 1973 Middle East war. In the years that followed, as U.S.-Soviet relations rapidly deteriorated, it is doubtful that leaders in either government ever regarded the statement of principles as more than a rhetorical artifact.[4]

Superpower crises would be made far less dangerous if the U.S.-Soviet relationship were substantially "denuclearized." That could not mean proscribing nuclear weapons, since the know-how necessary to build them cannot be outlawed, but—as the only feasible alternative—so restructuring American and Soviet forces as to make it clear that each retains nuclear weapons solely to deter their use by the other or by lesser powers. The way to bring about this condition would be, first, to halt the further deployment and testing of nuclear warheads and delivery systems (the "freeze"); second, to negotiate large reductions in the size of the superpower nuclear arsenals ("deep cuts"); and, third, for the United States to declare, as the Soviet Union already has, that it would not use nuclear weapons except in retaliation for their use by others ("no first use").[5]

There are strong strategic reasons for moving toward a posture of unambiguous deterrence (as distinguished from a posture in which nuclear weapons seem usable for "war fighting"). There are also strong political and psychological reasons. It would tell the military-industrial establishments of both sides that they will acquire no new nuclear weapons and that their current arsenal will not only diminish in number but, to the extent that they cannot be tested or replaced, will also become increasingly less useful for executing war-fighting scenarios that depend on high accuracy and reliability. A formal bilateral accord along these lines would give the two governments a means of resisting the inexorable pressure for "modernization"—for better and, usually, more weapons—that drives the nuclear arms race.

Such a posture might spill over in important ways to other areas of U.S.-Soviet relations and clear away some of the psychological and bureaucratic obstacles to a more satisfactory modus vivendi between the two superpowers. Under existing conditions, the ever-present possibility of nuclear war has made their relationship like none other in history. Never previously have two states lived with the continuing knowledge that each could almost instantly be destroyed by the other. That possibility and the measures each has independently taken to forestall it necessarily exacerbate a distrust that colors every aspect of the relationship, magnifying conflicts over other issues and giving rise at moments of tension to an upwardly spiraling hostility. So long as nuclear weapons exist, that distrust is unlikely to vanish altogether. But the knowledge that neither possesses nuclear forces configured or deployed so as to facilitate an attack on the other might well have a confidence-building and perhaps even transforming effect in other domains. In particular, it might defuse the processes that cause nearly any incident or dispute to ratchet up the overall level of tension between Washington and Moscow.

So drastic a change in the declared role of nuclear weapons, and the deployment changes necessary to give it meaning, will not be easily achieved, however. Nuclear weapons have played an important role in the military strategies of both governments. Especially is that true for Washington, which ever since the late 1940s has assumed that the threat of escalation across the nuclear threshold would compensate for a perceived deficit in conventional forces. Recent U.S. planning has been schizophrenic. In an effort to raise the nuclear threshold, the Army has been developing new, "smart" conventional weapons to accomplish many of the tasks once relegated to nuclear weapons. Yet, simultaneously, all three services have been developing small, highly accurate nuclear missiles whose deployment would be extraordinarily difficult to verify. Within a few years it may be impossible without intrusive on-site inspection to monitor deployments of certain systems. And at that point even a freeze on reliability testing might not offer enough confidence to build a political base for reductions that would unambiguously signal a departure from today's potentially hair-trigger, nuclear war-fighting strategies.

Fundamental change in the nuclear realm is a desirable but not a necessary prerequisite to better relations between the United States and the Soviet Union. What is needed most of all is greater realism about the world beyond their borders. Each government tends to blame the other for setbacks suffered by its clients. Both underestimate the degree to which fragile, unrepresentative governments that depend on coercion to remain in power and do not enjoy support among their own populations are susceptible to being turned out of office regardless of the extent to which their opponents receive support from outside. And—what makes the problem so difficult—both too readily assume that their own security is directly affected by political change in client states.

For the Soviet Union that assumption is not entirely unreasonable. Any event that calls into question the legitimacy of a self-avowed Marxist-Leninist regime anywhere may ultimately be seen as calling into question the legitimacy of all such regimes, because none rests on a base of freely given consent. Moscow is thus justifiably nervous at the prospect of opposition to the Castro regime in Havana. It is even more nervous at the prospect of opposition to the puppet government it installed in Afghanistan, given the ethnic and linguistic identity between the Afghans and large numbers of Muslims in the Central Asian areas of the Soviet Union. But what causes the Soviet leaders the most anxiety is any development that might appear to threaten the communist regimes of Eastern Europe. There the Soviet regime's own legitimacy is truly on the line. Revolutionary currents among Afghans or even Cubans can be rationalized away, but if Polish workers succeed in establishing alternate centers of political power, such as Solidarity in its heyday, why should not Soviet workers attempt to do the same?

The United States faces no such potential threats to its domestic constitutional order. Political upheaval in Central America or even in Mexico might cause Americans to question their government's policies, but not its legitimacy. Nor, because of its fortunate geographical circumstances, does the United States face any severe military threats from its southern neighbors. In the worst (and least likely) case, even if one were to become the site of a Soviet base, that would not alter the global strategic balance. Compared to the costs that Moscow would incur in maintaining such a base at the end of long and vulerable lines of communication, the cost to the United States of destroying it would not be large.[6]

Geography and politics thus make for substantial asymmetry between the Soviet and American situations. The Soviet regime would be threatened by political change near its borders—especially in Eastern Europe—while its American counterpart would not. The operational consequences of this reality are significant. It means that Washington has an interest in persuading Moscow that it poses no military threat to the communist regimes in Eastern Europe and that, in the event of an upheaval there, the West would not intervene except with words or, if Soviet repressive actions were distasteful enough, perhaps with boycotts and other economic instruments.

That, also, is why the West has such a strong interest in breathing new life into the semimoribund NATO–Warsaw Pact negotiations for mutual and balanced force reductions, and in reaching agreement on confidence-building measures to reduce each side's fear that the other is planning a surprise attack. Clearly such an arms control regime would help diminish Western fears of a Soviet attack. But because Europe's real tinderbox will continue to lie east of the Elbe River, its greatest utility would lie in reassuring Moscow and Pankow, rather than Washington or Bonn.

By easing Moscow's paranoia, these measures might make possible a détente (i.e., a relaxation of tensions) between the superpowers elsewhere as well. They should not be seen as a panacea, however. Indeed, Moscow might agree to none of them, for such measures would make too plain to the peoples of Eastern Europe that the West poses no military threat. Thus they would undermine the main Soviet rationale for stationing large military forces there. Those forces have been used thus far not to fight the West but to assure the perpetuation of Soviet-style communist orthodoxy, and the conservative leadership in the Kremlin might fear to do without them. The West will never know that, however, unless it puts Moscow to the test.

What about the other side of the equation—Soviet promotion of instability among American clients? It would be perverse to argue that because such instability does not directly threaten U.S. vital interests it should be tolerated, whereas the West should sign a self-denying pledge regarding Eastern Europe. Clearly Washington should seek to reach an agreement with Moscow providing for mutual nonintervention in the internal conflicts of Third World states. But defining the limits of permissible activities—and doing so with sufficient specificity to provide concrete "rules of engagement" whose violation will be universally apparent—will be formidably difficult. It was the absence of specificity that made the 1972 U.S.-Soviet "Basic Principles" agreement not only useless but —because the U.S. side believed itself betrayed when Moscow failed to provide warning of the Egyptian attack on Israel a year later—actually harmful.[7]

Negotiators seeking agreement on the limits of permissible intervention should recognize that to curtail assistance to insurgents, it is probably necessary to curb assistance to incumbent governments as well. Such evenhandedness is not required by international law, which allows assistance to recognized sovereign governments while forbidding aid to insurgents. But it is unlikely that Moscow and Havana would agree to cease supporting Central American revolutionaries if Washington goes on giving military assistance to Central American governments. In Afghanistan, the situation is reversed. There Moscow props up the incumbents, while Washington aids the insurgents. The United States is unlikely to agree to stop its aid unless the U.S.S.R. agrees to withdraw its troops.

It may well be that it is impossible to formulate mutually acceptable rules of engagement on a global level, and that regional rules should be sought instead. One potentially promising example is the network of treaties put forward by the so-called Contadora Group proscribing intervention in Central America and governing relations among the states of the region.[8] There, it should be noted, agreement might be made easier by the fact that the problem of intervention is not simply one of the United States supporting governments against communist-supported insurgents; in Nicaragua, as in Afghanistan, the alignments are reversed.

Precisely because multilateral approaches to nonintervention tend to place insurgents and incumbents on the same level, American policymakers usually dismiss them as equating thieves and householders—never mind that many of the householders have been brutal regimes with few claims to popular loyalty. Washington has often rationalized its rejection of multilateral solutions, as the Reagan administration did with the Contadora proposals in 1984, by asserting that the communist side would cheat anyway—despite the fact that when such approaches have been tried (the Geneva Accords on Indochina of 1954 and the Paris peace agreement on Vietnam of 1973 are leading examples), the U.S. record of compliance has been no better. And U.S. officials have tended to assume that without fetters the American side would be advantaged, since the United States could put more military assistance (matériel, advisers and trainers, even combat troops) into the hands of incumbents than the Soviets and their allies could get into the hands of insurgents. The Pentagon seems to worry that in some places its clients cannot survive without prolific U.S. support even if the insurgents are receiving no aid of any significance.

These are shortsighted calculations. They symbolize the gap between American official rhetoric, which has traditionally expressed a hope for cooperative, negotiated solutions to political-military problems, and what has actually been a strong U.S. preference for unilateral, imposed outcomes. However, so long as they are sufficiently concrete, with adequate mechanisms for monitoring and enforcement, multilateral agreements limiting intervention (or, in some instances, bilateral U.S.-Soviet agreements) offer considerable advantages. Most obvious, but also potentially most important, they remove an issue from the arena of East-West competition. At some moments, with some issues, that in itself might be a step toward preventing war. And even when the issues come nowhere near involving vital interests, unilateral interventions affect other aspects of U.S.-Soviet relations. Arms control becomes much more difficult. SALT II nearly foundered on the rocks of Soviet-Cuban intervention in Angola and Ethiopia; the treaty died altogether when Soviet forces moved into Afghanistan.

Moreover, interventions breed counterinterventions and create a climate in which people keep score. In campaigning for reelection in 1984, Reagan made much of the fact that no nation had "fallen to communism" on his watch, and he successfully portrayed the overwhelming assault by U.S. forces on a tiny Cuban garrison on Grenada as a triumph. It would be instructive to know what his Soviet counterparts, campaigning within the closed political circles of the Kremlin, claimed as their triumphs. (The accepted wisdom is that the Soviet list would include Angola and Mozambique, Ethiopia, Nicaragua, and perhaps even Iran. But from Moscow these triumphs may look more encumbered with costs than blessed with benefits. Even leaving aside Iran, whose regime's posture is overtly anti-Soviet, the other governments would be far less inclined to follow Moscow's line closely were it not for U.S. pressure against them.)

Such claims and counterclaims are part of the everyday business of Soviet-American relations today, a quite insidious process in which the energies and talents of the leading members of the political establishments of both superpowers are focused on the competition between them, and in which vast resources are annually poured into preparing for war. Americans and Soviets have come to accept it as utterly normal that their two governments should spend more than half a trillion dollars annually on military forces. The main purpose of those preparations for war is to deter the other side from attacking, and it is reasonable to suppose that the professionalism and dedication of the two military establishments will enable them to succeed in that goal.

But there is always a chance that they will fail. And, in any case, a by-product of their constantly growing capabilities is a propensity to make war, or to help proxies make war, beyond the European borderlands that divide East from West. As the United States discovered in Vietnam and the Soviet Union has found in Afghanistan, those efforts can severely tax the superpowers themselves, but they are vastly more costly to the Third World societies that are the arenas for their competition. That is not to suggest that the Third World would not be the scene of painful conflicts in the absence of superpower intervention; nor is it to say that intervention has never had a damping effect. But for every conflict eased, others there have been exacerbated, often at a fearful price.

The United States is not really estranged from the Soviet Union. Not even in the days of high détente during the Nixon administration did the U.S.-Soviet relationship have a quality whose diminution (indeed, disintegration) should be labeled an estrangement. But U.S.-Soviet business as usual, especially as conducted by the Reagan administration, has contributed in fundamental ways to the three types of estrangement enumerated earlier in this essay.

The first of these estrangements is between the rich and the poor nations. Washington's obsession with Moscow has reduced the absolute amount of economic resources available for economic development. Military expenditures, of course, compete with development assistance funds in the budgets of all the nations of the West—and no doubt in the budgets of the communist states as well. There is no reason to think that a dollar (or a ruble) saved on arms would go toward development on anything like a one-for-one basis. The Soviets, for their part, have consistently—and cynically—made clear that they feel no obligation to contribute economic aid to poor countries; that, they say, is the responsibility of the former colonial (and present "imperialist") powers of the West. Yet any Soviet contribution aside, it is painfully obvious that if even a small fraction of the world's military expenditures were to be converted to development purposes, the effect would be significant.

In part because of its large defense outlays, at no time over the past two decades has the United States contributed anything like as large a share of its wealth toward economic assistance as most of the other nations of the West. But in recent years the combination of gigantic government deficits and the Reagan administration's insistence nevertheless on very large increases in an already swollen defense budget has made foreign aid more than usually vulnerable to cost paring, and has made the U.S. record as a donor seem even more dismal. Moreover, to a much greater extent than its predecessors, the Reagan administration has used economic assistance to support its global anti-Soviet political agenda; funds have gone disproportionately to a few states, including El Salvador, Israel, Egypt, and the Philippines.

U.S.-Soviet competition has not only left fewer resources for development assistance. It has also spilled over into, and fueled, regional arms races in the Third World. Particularly in the Middle East the superpowers have been assiduous merchants of arms. Their sales—like those of other producers, such as the United Kingdom and France—have clearly helped their own economies. But a powerful motivation for both the United States and the U.S.S.R. has been the cultivation of clients. This is not to say that Third World regimes would not have sought (and in some measure acquired) arms regardless, but a different posture on the part of Washington and Moscow—one that discouraged the introduction of new and more sophisticated weapons into regional conflicts—might have had a restraining effect. And it might have slowed an alarming process of militarization that has distorted the priorities of many Third World regimes.

Estrangement exists not merely between the United States and much of the Third World, but between the richer and the poorer nations in general. It is reflected in the absence today of anything resembling the North-South dialogue that was so much a part of the atmosphere of international relations during the 1970s. Highly rhetorical (especially on the southern side), the dialogue was deficient in many respects. But it represented a commitment by the advanced, industrialized, market-economy states to the principle of taking coordinated steps to hasten development in the South.

During the 1980s the climate has been different. From Washington, London, Bonn, and other capitals have come injunctions that the South should solve its own problems. What has changed most of all, however, has been the stand of the United States. No longer is America perceived as committed to the goal of alleviating global poverty, and the evident absence of such a commitment—and therefore of any demonstrated leadership—on the part of the world's largest national economy gives others an excuse for defection too.

The estrangement that exists among the industrialized democracies—disagreements over issues such as coordinating macroeconomic policies or coping with Third World debt—has also been exacerbated by Washington's approach to relations with the Soviet Union. Since 1980 the United States has vastly increased its own military expenditures and pressured its allies, including Japan, to increase theirs as well. Most have complied reluctantly and, by the reckoning of the Reagan administration, insufficiently. All have been bruised by the competition between military expenditures and other national purposes. Moreover, because the other nations of the West have recovered from the recession of the early 1980s so much less rapidly than the United States, they have felt more acutely the traditional trade-off between butter and guns. The West has also been divided by differences over economic relations with the East. During the last half-dozen years two U.S. administrations have sought to impose a variety of economic penalties on the Soviet Union for its alleged political misdeeds, while most European states and Japan have been disinclined to do so.

On top of these economic differences, there are political-military ones as well. The allies part company with the Reagan administration over its assessment of the worldwide nature of the communist threat and its contention that revolutions in Central America or Africa, or terrorist threats to Israel, all stem in large measure from the same Soviet source. Some governments—and large portions of the European public—differ with Washington over questions of nuclear arms control, particularly over the deployment of intermediate-range missiles in Europe. And there are likely to be very serious disagreements if the Reagan administration's so-called Strategic Defense Initiative—and the race for space supremacy that it implies—ever gathers substantial momentum. All of these divisions will be enlarged if the political pendulum swings and left-wing parties, radicalized by opposition and by generational change, return to power in certain capitals of Western Europe.

Behind such disagreements lie larger ones over the appropriate role of carrots and sticks in East-West relations—including the stick of military forces. For the moment, with conservative governments in office in much of Europe, these differences are manageable. But one should not forget that NATO has a long history of public quarrels; the United States does not, and could not, force its allies to display the kind of harmony the Soviet Union imposes on the Warsaw Pact. But, particularly if the Reagan administration passes up the opportunity for significant arms control agreements with Moscow, it may be that the European-American disputes of the late 1980s will be even more divisive than previous ones have been.

The estrangement of many governments from the entire idea of international cooperation has also grown deeper as the Soviet-American confrontation has grown more bitter. The U.S.-Soviet rivalry has divided and made less productive most of the United Nations family of international agencies almost from the start. But a number of trends have compounded this phenomenon over the last decade. One is the persistence of the Arab-Israeli conflict and the growing realization that Israel is likely never to vacate the West Bank territory it occupied in 1967. A second is the continued existence of apartheid in South Africa. The international conflict in the Middle East and the domestic struggle in South Africa have often seemed to dominate the deliberations and stultify the work of the entire spectrum of international organizations. Moreover, because the West—and, in particular, the United States—has been perceived as the main support of both Israel and the white regime in South Africa, the Soviet Union and its allies have found it easy to pose as the primary proponents of the claims of both the Palestinians and South African blacks. Therefore, because all these fault lines come together and reinforce one another, Moscow has been able to rally Third World support for its positions on many other issues.

Another trend is the growing alienation of many of the new states that achieved independence from colonial rule during the late 1950s and early 1960s. They have come to realize that the heady optimism of their early years was misplaced, and that, instead of achieving self-sustaining growth, they may be doomed to permanent poverty. That realization was brought on by the changes in the international economy triggered by the rise in oil prices following the Arab-Israeli war of 1973. But rather than hold the Arab oil producers responsible for their troubles, the Third World states that have suffered the most have tended to blame the West—most of all the United States—and to vote against Washington when they could. During the Carter administration, skillful diplomacy at the United Nations worked against these tendencies. Since 1981, however, the intensely ideological approach of the Reagan administration has made it seem contemptuous of and hostile toward the Third World in general and the United Nations in particular.

To lay these multiple estrangements at Washington's doorstep, and to blame the United States alone, might seem far-fetched. Estrangement is a two-way street. A wealthy and powerful state in a world of poor states that perceive themselves as weak will inevitably be the object of widespread resentment. It has long been the case that other governments find it useful in their domestic politics if they can blame the United States for their own failures. Moreover, what does it matter? Interstate relationships are based on interest, not sentiment. Governments (and peoples) ask what the United States has done for them lately—and usually go on to ask for more. But that is precisely the point. From Washington, these days, they often get less. The Reagan administration's obsession with the Soviet Union has diverted resources from more productive uses both at home and abroad, and it has made more difficult the taking of initiatives commensurate with America's size and wealth. Preoccupied with the loss of hegemony, Reagan and his associates have often asserted their point of view loudly, but by and large they have failed to exert leadership.

Moreover, the administration has made it alarmingly easy for publics in the Third World and even in Western Europe and Japan to equate the United States with the Soviet Union. That is perhaps the administration's greatest failing in foreign policy. To many foreign observers, the superpowers now appear equally unwilling to tolerate ideological diversity among their neighbors and equally inclined to use military power to enforce conformity; each seems to be motivated by the narrowest conceptions of self-interest, and to regard the projection of force as its principal instrument of diplomacy.

For the United States, this equation may give rise to the most damaging estrangement of all, between it and its natural constituents abroad—those persons in many societies who might be called modernizing democrats. They are the governing majority in some countries, a beleaguered, suppressed minority in others. In Argentina, Brazil, and Uruguay they have recently come to power after years of military rule. In Chile and the Philippines they are a potent opposition to the reigning dictatorships. In Iran they tried, and failed, to offer a middle way between the autocracy of the Shah and the theocracy of the ayatollahs. Their political orientation may be right or left of what Americans would label as centrist, but they share a dedication to the liberal values of toleration and due process that are embedded in the U.S. Constitution and, by and large, characterize American domestic politics.

The Carter administration targeted its human rights policy on such persons. Its execution was flawed. Inconsistencies abounded. Claims of national security often needlessly took precedence over claims for human rights. Yet in conception it was sound policy to criticize publicly and withhold resources from governments guilty of egregious violations of human rights. Many persons in many countries escaped execution or torture and many more political prisoners were released because of America's explicitly stated concern. Despite its inconsistencies, Carter's approach was universally applauded by human rights activists in repressive countries. Better to keep the heat publicly on some governments than on none, or than to apply the heat cynically—as the Reagan administration has done—only to avowed enemies such as Sandinist Nicaragua.

The victims of human rights abuses are to a considerable extent these modernizing democrats. In the past they have looked to the United States for support and, in many instances, for inspiration. Americans have a twofold interest in their well-being. First is the selfish assumption that their societies will be more hospitable to Americans—as tourists, as traders, and (if Americans choose) as partners in constructing a more benign world order—if their countries are relatively open and devoted to due process and the rule of law. Second is an interest in the development of human potential for its own sake, and an assumption that such development is vastly more difficult in societies that are closed and repressive, no matter whether the repression is of the totalitarian or the authoritarian variety.

The supposition that human development is an American interest rests on another assumption: that it is possible, under compassionate and imaginative leadership, for the welfare of other societies increasingly to become a concern of citizens of the advanced, industrialized countries. That is already the case in the northern European democracies, in particular the Netherlands and Sweden, countries that have gone far toward solving their own most severe social problems. For young people who see few moral challenges at home, causes like ending apartheid in South Africa or relieving widespread misery in the Sahel or the Horn may have become something of a substitute. The United States has more social problems within its own borders, and more intractable ones, than do the highly homogeneous, relatively affluent nations of, say, Scandinavia. But it is possible to imagine that in America, too, the quest for equity at home will (as began to occur in the 1960s) give rise to a search for ways to enhance equity abroad.

These are paths to reconciliation. They begin with the Soviet Union, for only a drastic deescalation of the U.S.-Soviet conflict can liberate American policymakers to search for cooperative solutions to many of the most vexing problems that beset every one of the world's nations. Quite clearly, to break out of a forty-year pattern of spiraling mistrust will not be easy. For both Washington and Moscow the present relationship is strenuous but probably not lethal, and (for most of the citizens of both countries) not even acutely uncomfortable. To depart from the familiar path of "negotiation from strength" (or no negotiations at all) will require a vision of a possible cooperative future certainly not evident in the first administration of Ronald Reagan and perhaps not likely to emerge in the second. Planned weapons deployments will make the task more difficult for his successor, but, given sufficient political will, not impossible. Meanwhile, the delay will cost many hundreds of billions of additional dollars wasted on armaments, and no doubt some wasted lives as well.

Unless the focus of debate is widened, however, 1988 will offer no greater opportunity to break out of this pattern than 1984, when the two major U.S. presidential candidates vied to present an image of maximum "toughness" and seemed to put forward only slightly different versions of the same foreign policy. I have tried in the foregoing pages to put forward a quite substantially different vision. If the United States seeks to end the estrangement between itself and much of the rest of the world, it must start by getting a clearer notion of the price it pays in lost opportunities by continuing with business as usual.

Notes

1. UNGAR

1. Barry Rubin, *Paved with Good Intentions: The American Experience and Iran* (New York: Oxford University Press, 1980), pp. 200–201; and Gary Sick, *All Fall Down: America's Tragic Encounter with Iran* (New York: Random House, 1985), pp. 30 and 62.

2. See Daniel Patrick Moynihan (with Suzanne Weaver), *A Dangerous Place* (Boston: Atlantic–Little, Brown, 1978).

3. "American Values: Change and Stability," a conversation with Daniel Yankelovich, *Public Opinion*, December/January 1984, p. 5.

4. Christoph Bertram, "When America was a Superpower," *Washington Post*, November 25, 1984, p. C-7. For McLuhan on the global village, see *War and Peace in the Global Village*, written with Quentin Fiore (New York: McGraw-Hill, 1968).

5. Jeane J. Kirkpatrick, *The Reagan Phenomenon—and Other Speeches on Foreign Policy* (Washington: American Enterprise Institute for Public Policy Research, 1983), p. 30; Moynihan, op. cit., p. 236; and Michael Ledeen, *Grave New World* (New York: Oxford University Press, 1985), pp. 25 and 28.

6. Jonathan Kwitny, *Endless Enemies: The Making of an Unfriendly World* (New York: Congdon & Weed, 1984), p. 394; T. D. Allman, *Unmanifest Destiny* (New York: Doubleday, 1984), p. 376; and I. M. Destler, Leslie H. Gelb, and Anthony Lake, *Our Own Worst Enemy: The Unmaking of American Foreign Policy* (New York: Simon & Schuster, 1984), p. 12.

7. Fouad Ajami, "The End of the Affair: An American Tragedy in the Arab World," *Harper's*, June 1984, p. 63.

8. Barry Sussman, "El Salvador is Not in Louisiana," *Washington Post*, January 2, 1983, p. B-5.

9. *Strength through Wisdom: A Critique of U.S. Capability, A Report to the President from the President's Commission on Foreign Language and International Studies* (Washington: U.S. Government Printing Office, November 1979), pp. 7–8 ff. See also Paul Simon, *The Tongue-Tied American: Confronting the Foreign Language Crisis* (New York: Continuum, 1980).

10. Howard J. Wiarda, *Ethnocentrism in Foreign Policy: Can We Understand the Third World?* (Washington: American Enterprise Institute for Public Policy Research, 1985), p. 1.

11. See, for example, W. T. Stead, *The Americanization of the World, or*

The Trend of the Twentieth Century (New York and London: Horace Markley, 1901).

12. Wiarda, op. cit., pp. 4 and 29.

13. Georgi A. Arbatov, "Relations Between the United States and the Soviet Union—Accuracy of U.S. Perceptions," in *Toward Nuclear Disarmament and Global Security: A Search for Alternatives,* Burns H. Weston, ed. (Boulder, Colo.: Westview Press, 1984), p. 280.

14. Ronald Steel, "Behaving Like a Great Power," *Vanity Fair,* February 1984, pp. 36–37. See also Ronald Steel, *Pax Americana* (New York: Viking, 1967).

2. DALLEK

1. For the Leahy and LaFeber statements, see Walter LeFeber, *America, Russia and the Cold War, 1945–1980,* 4th ed. (New York: John Wiley, 1980), pp. 24 and 27. For the popularity of Willkie's *One World* and the poll data, see Robert Dallek, *The American Style of Foreign Policy: Cultural Politics and Foreign Affairs* (New York: Knopf, 1983), pp. 132–33.

2. For the comments about the Soviet Union, see Dallek, op. cit., pp. 139–40; and Robert Dallek, *Franklin D. Roosevelt and American Foreign Policy, 1932–1945* (New York: Oxford University Press, 1979), p. 520. For the Reagan quote, see *New York Times,* May 18, 1981, p. B-7. On China, Germany, and Japan, see Dallek, *The American Style of Foreign Policy,* pp. 143–50.

3. Vandenberg is quoted in Eric F. Goldman, *The Crucial Decade—And After: America, 1945–1960* (New York: Vintage, 1960), pp. 29–30. Luce is quoted in Dallek, *The American Style of Foreign Policy,* pp. 141–42.

4. On the Bretton Woods agreements, see Lloyd C. Gardner, *Economic Aspects of New Deal Diplomacy* (Boston: Beacon Press, 1971), pp. 285–90, and John Morton Blum, *V Was for Victory: Politics and American Culture during World War II* (New York: Harcourt Brace Jovanovich, 1976), pp. 307–9.

5. For the Bruner quote and the state of American domestic feeling, see Dallek, *The American Style of Foreign Policy,* pp. 132–38. For feelings about Bretton Woods, see the references in note 4, and Gabriel Kolko, *The Politics of War: United States Foreign Policy, 1943–1945* (New York: Random House, 1968), Chap. 11 and pp. 255–66 in particular.

6. For FDR's wartime diplomacy, see Part Four of Dallek, *FDR and American Foreign Policy.* The quotes are on pp. 422, 466, and 522.

7. On Truman, see Robert J. Donovan, *Conflict and Crisis: The Presidency of Harry S. Truman, 1945–1948* (New York: W. W. Norton, 1977), pp. 12–13, and Chap. 4.

8. On the China policy, see ibid., pp. 246–48.

9. For Truman's policy on atomic energy, see ibid., pp. 132, 134–35, 156–57, 203–6. Also see Gregg Herken, *The Winning Weapon: The Atomic Bomb in the Cold War, 1945–1950* (New York: Knopf, 1980), Chaps. 8–9. The quotes are on pp. 160, 170, and 174–75.

10. For Churchill's and Wallace's speeches, see LaFeber, op. cit., pp. 39–40, 43–45. For the liberal reaction to the Churchill speech, see John L. Gaddis, *The United States and the Origins of the Cold War, 1941–1947* (New York: Columbia University Press, 1972), p. 309.

11. On the Truman Doctrine, see Gaddis, op. cit., who quotes Truman, pp. 348–52, and LaFeber, op. cit., pp. 52–57. Also see Joseph M. Jones, *The Fifteen Weeks* (New York: Viking, 1955), pp. 138–42; and Dean Acheson, *Present at the Creation* (New York: W. W. Norton, 1969), p. 219, for Acheson's remarks. Dallek, *The American Style of Foreign Policy,* quotes Lippmann on p. 156. Adam Ulam, *The Rivals: America and Russia Since World War II* (New York: Viking, 1971), pp. 121–26.

12. Donovan, op. cit., pp. 287–91. Ulam, op. cit., pp. 126–30

13. Ronald Steel, *Walter Lippmann and the American Century* (Boston: Atlantic—Little, Brown, 1980), pp. 440–49.

3. DONOVAN

1. Figures supplied by the U.S. Department of Defense.

2. Bruce Cumings, *The Origins of the Korean War: Liberation and the Emergence of Separate Regimes, 1945–1947* (Princeton, N.J.: Princeton University Press, 1981), p. 102.

3. Hugh Brogan, "Drawing the Line in Korea," *Times Literary Supplement* (London), August 26, 1983, p. 902.

4. Records Office, London, Cabinet 85 (50). Conclusions of a Meeting of the Cabinet held at 10 Downing Street, December 12, 1950, pp. 224–25.

4. McHENRY

1. Willard Range, *Franklin D. Roosevelt's World Order* (Athens, Ga.: University of Georgia Press, 1959), p. 103.

2. Ibid., pp. 102–4.

3. *Victory and the Threshold of Peace: The Public Papers and Addresses of Franklin D. Roosevelt, 1944–45,* Samuel I. Rosenman, ed. (New York: Harper & Brothers, 1950), p. 69.

4. Ibid., pp. 70, 563–64.

5. Tony Smith, "Patterns in the Transfer of Power: A Comparative Study of French and British Decolonization," in Prosser Gifford and William Roger Louis, eds., *The Transfer of Power in Africa: Decolonization 1940–1960* (New Haven: Yale University Press, 1982).

6. John F. Kennedy, *The Strategy of Peace,* Allan Nevins, ed. (New York: Harper & Brothers, 1960), p. 67.

7. Philip W. Bonsal, *Cuba, Castro and the United States* (Pittsburgh: University of Pittsburgh Press, 1971), p. 21.

8. *Department of State Bulletin,* vol. XXXIV, June 18, 1956, pp. 999–1000.

5. HEHIR

1. The following synthesizes more extensive analysis which can be found in Stanley Hoffmann, *Gulliver's Troubles: or the Setting of American Foreign Policy* (New York: McGraw-Hill, 1968).

2. Henry A. Kissinger, "Central Issues of American Foreign Policy," in Kermit Gordon, ed., *Agenda for the Nation* (Washington: The Brookings Institution, 1968), p. 589.

3. Hoffmann, op. cit., pp. 33–43.

4. Michael Mandelbaum, *The Nuclear Question: The United States and*

Nuclear Weapons, 1946–1976 (Cambridge: Cambridge University Press, 1979), p. 74.

5. Lawrence Freedman, *The Evolution of Nuclear Strategy* (New York: St. Martin's, 1981), p. 228.

6. Ibid.

7. Bernard Brodie, *The Absolute Weapon* (New York: Harcourt Brace, 1946), p. 76.

8. Robert S. McNamara, *The Essence of Security: Reflections in Office* (New York: Harper and Row, 1968), p. 52.

9. Henry A. Kissinger, *Nuclear Weapons and Foreign Policy* (New York: Harper and Row, 1961).

10. Mandelbaum, op. cit., p. 191.

11. See, for example, Graham T. Allison, *The Essence of Decision* (Boston: Little, Brown, 1971); Roger Hilsman, *To Move a Nation* (New York: Dell, 1964), pp. 413–540; James A. Nathan, "His Finest Hour Now," *World Politics* 27 (1975), pp. 256–81; Peter Rodman, "The Missiles of October: Twenty Years Later," *Commentary*, October 1982, pp. 39–45.

12. For contrasting accounts of what was the decisive influence in the crisis and its implications for policy, see Albert and Roberta Wohlstetter, "Controlling the Risks in Cuba," *Adelphi Paper* #17 (London: Institute for International and Strategic Studies, 1965); McGeorge Bundy, "The Unimpressive Record of Nuclear Diplomacy," in Gwyn Prins, ed., *The Nuclear Crisis Reader* (New York: Random House, 1984), p. 50; Rodman, op. cit.

13. Raymond Aron, *Peace and War: A Theory of International Relations* (New York: Doubleday, 1966), pp. 441 ff.

14. Recent accounts include Stanley Karnow, *Vietnam: A History* (New York: Viking, 1983); Guenter Lewy, *America in Vietnam* (New York: Oxford University Press, 1978); Richard M. Nixon, *No More Vietnams* (New York: Arbor House, 1985).

15. Henry A. Kissinger, *The White House Years* (Boston: Little, Brown, 1979,) p. 57.

16. Nathan, op. cit., p. 281

17. Neil Sheehan, "The Covert War and Tonkin Gulf: February-August 1964," in Neil Sheehan et al., *The Pentagon Papers as Published by the* New York Times (New York: Bantam, 1971), p. 255.

18. Arthur M. Schlesinger, Jr., *A Thousand Days: John F. Kennedy in the White House* (Boston: Houghton, Mifflin, 1965), p. 303; David Halberstam, *The Best and the Brightest* (New York: Random House, 1969), p. 122.

19. Stanley Hoffmann, "Vietnam and American Foreign Policy," in Richard Falk, ed., *The Vietnam War and International Law*, vol. II (Princeton, N.J.: Princeton University Press, 1969), pp. 1148–49.

20. The history of the air war can be found in *The Pentagon Papers: The Defense Department History of United States Decisionmaking on Vietnam*, vol. IV (Boston: Beacon Press, 1971).

21. Halberstam, op. cit., p. 68.

22. Quoted in Lewy, op. cit., p. 414.

23. Leslie H. Gelb, "The Essential Domino: American Politics and Vietnam," *Foreign Affairs* (April 1972), pp. 459–75.

24. *Pentagon Papers,* vol. IV, p. 172.

25. For a detailed review of the concept of limited war, see Robert Osgood, "The Post-War Strategy of Limited War: Before, During and After Vietnam," in Laurence Martin, ed., *Strategic Thought in the Nuclear Age* (Baltimore: Johns Hopkins University Press, 1979), pp. 93–130.

26. W. Stein, ed., *Nuclear Weapons and the Christian Conscience* (London: Merlin, 1961).

27. Paul Ramsey, *The Just War: Force and Political Responsibility* (New York: Scribners, 1968).

28. The National Conference of Catholic Bishops, *The Challenge of Peace: God's Promise and Our Response* (Washington: The U.S. Catholic Conference, 1983).

29. See Stanley Hoffmann, *Duties Beyond Borders* (Syracuse, N.Y.: Syracuse University Press, 1982), for a good review of the policy literature; Kenneth W. Thompson, ed., *The Moral Imperatives of Human Rights: A World Survey* (Washington: University Press of America, 1980).

30. See J. Bryan Hehir, "Human Rights and National Interest," *Worldview* 25 (1982), pp. 18–21.

31. Anthony Lake, ed., *The Vietnam Legacy: The War, American Society and the Future of American Foreign Policy* (New York: New York University Press, 1976), p. xvii.

6. HODGSON

1. Daniel P. Moynihan, "Peace," address to Andrew Academy, January 1973, published in *The Public Interest,* Summer 1973, and in Daniel P. Moynihan, *Coping: On the Practice of Government* (New York: Random House, 1973).

2. Don Cook, *Charles de Gaulle: A Biography* (New York: Putnam, 1984), p. 354. I have corrected Mr. Cook's version of the French phrase used.

3. Data from Survey Research Center, University of Michigan. See also Albert H. Cantril, "The American People, Vietnam and the Presidency," paper given at the annual meeting of the American Political Science Association, September 1970; and Philip E. Converse and Howard Schuman, "Silent Majorities and the Vietnam War," *Scientific American,* June 1970.

4. Robert J. Donovan, *Tumultuous Years: The Presidency of Harry S Truman, 1949–1953* (New York: W. W. Norton, 1982), pp. 159–61; and Dean Acheson, *Present at the Creation* (New York: W. W. Norton, 1969), pp. 373–80.

5. See, among others, Stephen E. Ambrose, *Rise to Globalism: American Foreign Policy Since 1938* (New York: Penguin, 1983).

6. Andrew Hacker, *The End of the American Era* (New York: Atheneum, 1970), p. 230.

7. Richard Whalen, *Catch the Falling Flag* (Boston: Houghton Mifflin, 1972).

8. The Nixon Doctrine, foreshadowed by Richard M. Nixon in a 1967 *Foreign Affairs* article (Vol. 46, No. 1., October 1967), was outlined in more detail to the White House press corps during a July 1969 Pacific trip. See Henry Brandon, *The Retreat of American Power* (London: Bodley Head, 1972), pp. 79–83.

9. Interviews with the author in Bonn, 1977.

10. See the Organization for Economic Cooperation and Development's *Main Economic Indicators* series and the World Bank's annual *World Development Report* for comparative data on GNP and income.

7. THUROW

1. Roger Brinner and Nigel Gault, "U.S. Manufacturing Costs and International Competition," *Data Resources Review*, October 1983, p. 1.15.

2. U.S. Department of Commerce, *International Economic Indicators*, September 1984, p. 64. The annual growth rates are derived from the indices provided in these data.

3. Ibid.

4. Council of Economic Advisers, *Economic Report of the President*, February 1984 (Washington: U.S. Government Printing Office, 1984), p. 267.

5. Ibid.

6. Ibid., p. 221.

7. Council of Economic Advisers, *Economic Report of the President*, February 1985 (Washington: U.S. Government Printing Office, 1985), pp. 232–33.

8. International Monetary Fund, *International Financial Statistics*, March 1984, p. 264.

9. Andrew Pollack, "Japanese Lead in RAM Race," *New York Times*, February 2, 1984, p. D-2; "America's High-Tech Crisis: Why Silicon Valley Is Losing Its Edge," *Business Week*, March 11, 1985, p. 62.

10. "Thicker Than R2D2," *The Economist*, January 21, 1984, p. 63.

11. Data drawn from U.S. Department of Commerce, *Current Population Reports, Consumer Income, Money Income, and Poverty Status of Families and Persons in the United States: 1983*, Series P-60, no. 145, p. 18; and p. 19 in the 1968 edition of the same publication.

12. Ibid.

13. *Economic Report of the President*, 1984, pp. 221 and 249.

14. Mancur Olson, *The Rise and Decline of Nations: Economic Growth, Stagflation, and Social Rigidities* (New Haven: Yale University Press, 1982).

15. U.S. Department of Commerce, *International Economic Indicators*, September 1984, p. 15.

16. National Science Foundation, *National Pattern of Science and Technological Resources*, 1982, p. 33.

17. Edward B. Fiske, "American Students Score Average or Below in International Math Exams," *New York Times*, September 23, 1984, p. 30.

18. Morgan Guaranty Trust, *World Financial Markets*, February 1985, p. 16.

8. MAZRUI

1. See Benjamin Quartes, *Lincoln and the Negro* (New York: Oxford University Press, 1962); and Louis Ruchames, ed., *Racial Thought in America: From the Puritans to Abraham Lincoln* (New York: Grosset and Dunlap, 1969), pp. 380–82. Consult also Lerone Bennett, Jr., "Was Abe Lincoln a White Supremacist?" *Ebony*, February 1968, pp. 35–42; and Herbert Mitgang, "Was

Lincoln Just a Honkie?" *New York Times Magazine,* February 11, 1968, pp. 35, 100–107.

2. *Development Cooperation: Efforts and Policies of the Members of the Development Assistance Committee, 1984 Review,* (Paris: Organization for Economic Co-Operation and Development, 1984).

3. J. R. Barongo, "Understanding African Politics: The Political Economy Approach," *Nigerian Journal of Political Science* (December 1980), p. 68.

9. GEYELIN

1. Samuel Flagg Bemis, *The Latin American Policy of the United States* (New York: Harcourt, Brace, 1943), pp. 385 and 388.

2. Walter LaFeber, *John Quincy Adams and American Continental Empire* (Chicago: Quadrangle, 1965), p. 45.

3. In John Bartlett, *Familiar Quotations,* 15th ed. (Boston: Little, Brown, 1980), p. 264.

4. Source for figures: Overseas Development Council.

5. Louis M. Hacker and Benjamin B. Kendrick, *The United States Since 1865* (New York: F. S. Crofts, 1934), p. 351.

6. Mark Sullivan, *Our Times,* vol. I (New York: Charles Scribner's Sons, 1926), p. 47.

7. Ibid., p. 55–56.

8. Walter Lippmann, "Globalism and Isolationism," *New York Herald Tribune,* February 23, 1965, p. 26.

9. John F. Kennedy, "Transcript of Broadcast on NBC's Huntley-Brinkley Report," September 9, 1963, *Public Papers of the Presidents of the United States* (Washington: U.S. Government Printing Office, 1965), p. 659.

10. This decision is described by Henry A. Kissinger in *White House Years* (Boston: Little, Brown, 1979), pp. 220–22.

11. Jimmy Carter, "Address at Commencement Exercise at the University of Notre Dame," May 22, 1977, *Public Papers of the Presidents of the United States* (Washington: U.S. Government Printing Office, 1979), p. 956.

12. Jimmy Carter, in an interview with Frank Reynolds on ABC Evening News, December 31, 1979.

13. Henry Kissinger, *Years of Upheaval* (Boston: Little, Brown, 1982), p. 981.

14. Ronald Reagan, "Remarks at the Center's National Leadership Forum," April 6, 1984, *Weekly Compilation of Presidential Documents,* April 9, 1984, p. 496.

15. Jimmy Carter, op. cit., p. 956.

16. Kissinger, *White House Years,* p. 130.

17. Kissinger, *Years of Upheaval,* p. 980.

10. CHACE

1. MacLeish's piece of 1949 was reprinted in *Atlantic Monthly,* March 1980.

2. Jefferson quoted in Samuel Eliot Morison and Henry Steele Commager, *The Growth of the American Republic* (New York: Oxford University Press, 1940), p. 280.

3. Typical of European criticism of Carter, even when he tried to repair his mistakes, are these words from former German Chancellor Helmut Schmidt:

"President Carter . . . confronted his European allies with surprising 'lonely' decisions, taken without consultation. The situation was not eased when he made a number of subsequent corrections, since some of these were put into effect just as surprisingly." From "Saving the Western Alliance," *New York Review of Books,* May 31, 1984, p. 25.

4. See James Chace, "Is a Foreign Policy Consensus Possible?" *Foreign Affairs,* Fall 1978, pp. 1–16.

5. *New York Times,* May 10, 1982.

6. Stanley Hoffmann, "Reagan Abroad," *New York Review of Books,* February 4, 1982.

7. Quotation from President Cleveland's secretary of state, in James Chace, *Endless War* (New York: Vintage, 1984), p. 28.

8. James Chace, "Is a Foreign Policy Consensus Possible?" p. 2.

9. Source for figures: Organization for Economic Co-Operation and Development.

10. Council of Economic Advisers, *Economic Report of the President,* February 1984 (Washington: U.S. Government Printing Office, 1984).

11. See David Calleo, "U.S. Political-Economic Policy and Its International Economic Effects," Hudson Research Group, 1984.

12. Source for figures: U.S. Department of Commerce.

13. Source for figures: *Federal Reserve Bulletin.*

14. Dean Acheson, *Present at the Creation* (New York: W. W. Norton, 1969), p. 219.

15. *Wall Street Journal,* February 10, 1983.

16. Theodore Draper, "Falling Dominoes," *New York Review of Books,* October 27, 1983.

17. Ibid.

18. Ibid.

19. James Chace, "Deeper into the Mire," *New York Review of Books,* March 1, 1984; see also *The Report of the National Bipartisan Commission on Central America* (New York: Macmillan, 1984).

20. Ronald Reagan, November 11, 1982, news conference.

21. See Ze'ev Schiff and Ehud Yaari, *Israel's Lebanon War* (New York: Simon & Schuster, 1984).

22. *New York Times,* October 25, 1983.

23. Alexander M. Haig, *Caveat: Realism, Reagan, and Foreign Policy* (New York: Macmillan, 1984).

24. Henry A. Kissinger, *White House Years* (Boston: Little, Brown, 1979), p. 1114.

25. John E. Rielly, ed., "American Public Opinion and U.S. Foreign Policy 1983" (Chicago: Chicago Council on Foreign Relations, 1983).

26. Daniel Yankelovich and Larry Kaagan, "Assertive America," *Foreign Affairs: America and the World, 1980,* p. 711.

27. The Democratic Party Platform, July 1984.

28. Quoted by Michael Kramer, "Convention Snapshots," *New York,* July 30, 1984.

29. Barbara W. Tuchman, *The March of Folly* (New York: Knopf, 1984), p. 319.

30. Quoted in Frederick Merk, *Manifest Destiny and Mission in American History* (New York: Knopf, 1963), p. 9.

31. James Chace, "How Moral Can We Get?" *New York Times Magazine*, May 22, 1977.

32. Michel Crozier, "America, You Can't Afford Illusions," *Washington Post*, July 1, 1984, p. C-1.

11. FITZGERALD

1. Richard Nixon, "Address to the Nation on the Situation in Southeast Asia," April 30, 1970, *Public Papers of the Presidents of the United States* (Washington: U.S. Government Printing Office, 1971), p. 409.

2. Henry A. Kissinger, *White House Years* (Boston: Little, Brown, 1979), pp. 65 and 195.

3. John C. Fitzpatrick, ed., *The Writings of George Washington*, vol. 34 (Washington: U.S. Government Printing Office, 1940), p. 335.

4. Charles and Mary Beard, *America in Midpassage*, vol. I (New York: Macmillan, 1939), pp. 443 and 447.

5. For the material on Wilson, see Samuel Eliot Morison and Henry Steele Commager, *The Growth of the American Republic*, volume II (New York: Oxford University Press, 1950), Chapter XX.

6. Dean Acheson, *Present at the Creation* (New York: W. W. Norton, 1969), p. 219.

7. Frank N. Trager, *Why Viet Nam?* (New York: Praeger, 1966), p. 111.

8. On the overthrow of Mossadegh, see Frances FitzGerald, "Giving the Shah Everything He Wants," *Harper's*, November 1974, pp. 55–82; on the Guatemalan coup, see Stephen Schlesinger and Stephen Kinzer, *Bitter Fruit* (New York: Doubleday, 1982), pp. 159–204.

9. Kissinger, op. cit., pp. 65 and 195.

10. For a concise description of the origins of the doctrine during Nixon's July 1969 Pacific trip, and for excerpts from the key presidential briefings from which it was developed, see Tad Szulc, *The Illusion of Peace: Foreign Policy in the Nixon Years* (New York: Viking, 1978), pp. 124–28.

11. Jerry Falwell, *America Can Be Saved* (Murfreesboro, Tenn.: Sword of the Lord, 1979), p. 115.

12. For a general discussion of the influence of religious fundamentalism in American history, see Sidney E. Ahlstrom, *A Religious History of the American People* (New Haven: Yale University Press, 1972).

13. Richard Hofstadter, *The Paranoid Style in American Politics, and Other Essays* (Chicago: University of Chicago Press, 1979), p. 32.

14. Ronald Reagan, "The President's News Conference," March 6, 1981, *Public Papers of the Presidents of the United States* (Washington: U.S. Government Printing Office, 1982), p. 207.

15. Ronald Reagan, "Address Delivered Before a Joint Session of the Congress," April 27, 1983, *Weekly Compilation of Presidential Documents*, May 2, 1983, p. 614.

16. Ahlstrom, op. cit.

17. Lyndon Johnson, Address at Johns Hopkins University: "Peace With-

out Conquest," April 7, 1965, *Public Papers of the Presidents of the United States* (Washington: U.S. Government Printing Office, 1966), pp. 396–97.

18. Frances FitzGerald, "The Warrior Intellectuals," *Harper's*, May 1976, pp. 45–64.

19. William Watts and Lloyd A. Free, *State of the Nation* (Washington: Potomac Associates, 1973).

20. Henry A. Kissinger, *Years of Upheaval* (Boston: Little, Brown, 1982), p. 981.

21. Louis M. Hacker and Benjamin B. Kendrick, *The United States Since 1865* (New York: F. S. Crofts, 1934), p. 351.

22. Kissinger, *Years of Upheaval*, p. 981.

12. ULLMAN

1. See, for example, McGeorge Bundy, "After the Deluge, the Covenant," *Saturday Review/World*, August 24, 1975, pp. 18–20; or Robert Heilbroner, *An Inquiry into the Human Condition* (New York: W. W. Norton, 1975), esp. pp. 42–45.

2. For an acute and persuasive argument along these lines, see Coral Bell, *Negotiation from Strength* (New York: Knopf, 1963), a book too little known.

3. See Glenn T. Seaborg, *Kennedy, Khrushchev, and the Test Ban* (Berkeley: University of California Press, 1981).

4. For an analysis of the "Basic Principles Agreement," see Alexander L. George, *Managing U.S.-Soviet Rivalry: Problems of Crisis Prevention* (Boulder, Colo.: Westview, 1983), pp. 107–18. For Nixon's explanation, see Henry A. Kissinger, *White House Years* (Boston: Little, Brown, 1979), p. 1253.

5. For a lengthy statement of the arguments underlying this proposal, see Richard H. Ullman, "Denuclearizing International Politics," *Ethics* 95: 3 (April 1985), pp. 567–88.

6. See Joseph Cirincione and Leslie C. Hunter, "Military Threats, Actual and Potential," in Robert S. Leiken, ed., *Central America: Anatomy of Conflict* (New York: Pergamon Press, 1984), pp. 173–92.

7. See George, op. cit., pp. 365–76.

8. See the "Act of Contadora for Peace and Cooperation in Central America" (revised version), October 24, 1984, available from the General Secretariat of the Organization of American States, Washington, D.C.

Chronology

1940s

Aug. 14, 1941	Atlantic Charter, laying out principles for peace in postwar world and asserting right to national self-determination, issued by U.S. and Britain.
Dec. 7, 1941	Japanese planes attack U.S. naval base at Pearl Harbor, killing scores of enlisted men and destroying several naval vessels; U.S. declares war on Japan next day, marking entry into World War II; war declared on Germany and Italy Dec. 11.
Nov. 22–26, 1943	Franklin D. Roosevelt, Winston Churchill, and Chiang Kai-shek discuss war aims in Far East and surrender terms for Japan at Cairo Conference.
Nov. 28–Dec. 1, 1943	Roosevelt, Churchill, and Joseph Stalin meet in Tehran to discuss allied invasion of France and possible entry of U.S.S.R. into war against Japan.
July 1–22, 1944	Representatives of 44 nations gather at Bretton Woods, N.H., to design trade regime and institutional structure of postwar world economy; International Monetary Fund and World Bank established.
Feb. 4–11, 1945	Yalta Conference lays out terms for allied occupation of Germany and Stalin's entry into war against Japan; national boundaries and governmental structures for some East European countries determined.
April 12, 1945	Roosevelt dies; Harry S Truman becomes president.
April 21, 1945	Stalin concludes mutual-assistance pact with communist-led Lublin government of Poland.
April 30, 1945	Adolf Hitler commits suicide.
May 7, 1945	Germany surrenders unconditionally to Gen. Dwight D. Eisenhower at allied headquarters in Reims, France.
June 26, 1945	United Nations Charter signed by delegates from 50 countries gathered in San Francisco.
July 16, 1945	First atomic device successfully detonated in New Mexico test.

July 17–Aug. 2, 1945	Potsdam Conference, last of wartime summits, finalizes and implements terms of previous agreements on postwar Europe and termination of war in Pacific.
Aug. 6, 1945	First atomic bomb dropped on Hiroshima, Japan.
Aug. 9, 1945	Second atomic bomb dropped on Nagasaki, Japan.
Sept. 2, 1945	Japan formally surrenders to Gen. Douglas A. MacArthur aboard U.S.S. *Missouri* in Tokyo Bay.
Sept. 8, 1945	U.S. troops move into southern Korea as part of postwar occupation of territories formerly held by Japan.
Nov. 29, 1945	Assembly elected Nov. 11 proclaims Federal People's Republic of Yugoslavia with Marshall Tito as president.
March 5, 1946	In speech at Westminster College in Fulton, Mo., Churchill declares that "an iron curtain has descended across the continent" of Europe.
June 14, 1946	At first session of UN Commission on Atomic Energy, Bernard Baruch presents American proposal for international control of fissionable material and destruction of all stockpiles of atomic weapons.
July 4, 1946	Philippines gains independence from U.S. in accordance with timetable set earlier by Roosevelt and Congress.
Sept. 30, 1946	Guilty verdicts handed down for 22 former Nazi officials by international war crimes tribunal at Nuremburg.
Nov. 23, 1946	France launches military effort to retain Indochina colonies with bombing of Haiphong, Vietnam.
March 12, 1947	In major speech before joint session of Congress, Truman outlines ambitious new U.S. foreign commitment to counter spread of communism throughout the world, known as the "Truman Doctrine"; Truman calls upon Congress to appropriate $400 million in economic and military assistance to combat communist insurgency in Greece and political instability in Turkey.
June 5, 1947	In Harvard commencement address, Secretary of State George C. Marshall outlines program of financial aid for postwar European recovery, known as the Marshall Plan; Soviets reject Eastern bloc participation in the plan during July.
July 1947	Diplomat George F. Kennan advocates U.S. policy of "containment" toward U.S.S.R. in anonymous article in *Foreign Affairs*.

Aug. 15, 1947	India gains independence from Britain; Muslim areas remain dominions of British Crown, becoming Republic of Pakistan in March 1956.
Dec. 30, 1947	King Michael of Romania forced to abdicate by communist-dominated government.
Jan. 1, 1948	General Agreement on Tariffs and Trade (GATT), negotiated during 1947, comes into force, significantly lowering trade barriers among its 23 signatories.
Jan. 30, 1948	Indian independence movement leader Mahatma Gandhi assassinated by Hindu extremists.
Feb. 25, 1948	In bloodless coup in Czechoslovakia, President Eduard Benes forced to accept communist government under Klement Gottwald.
May 14, 1948	State of Israel proclaimed as British mandate in Palestine expires; neighboring Arab nations invade the new country the next day, commencing the "War of Independence" in which Israel significantly enlarges its land area and takes control of part of Jerusalem.
May 26, 1948	National Party prevails in parliamentary elections in South Africa, bringing the Afrikaner ethnic minority to political power; in coming years, National Party governments will construct body of laws and political and social institutions, known as apartheid, which enforce strict segregation of races.
June 28, 1948	Yugoslavia expelled from Cominform following series of disagreements with U.S.S.R.
July 25, 1948	U.S.-sponsored Berlin Airlift begins, delivering food and supplies to West Berlin, severed from greater West Germany by Soviet blockade.
Aug. 15, 1948	Republic of Korea (South) established.
Sept. 9, 1948	Democratic People's Republic of Korea (North) established.
Oct. 27, 1948	Voice of America, worldwide U.S. government radio broadcasting network, established by Congress.
Nov. 2, 1948	Truman defeats Republican Thomas E. Dewey in presidential election.
Dec. 27, 1948	Hungarian Cardinal Mindszenty arrested for outspoken criticism of communist government; death sentence later commuted to life imprisonment; in 1956, he will seek asylum in American Embassy.

Jan. 18, 1949	Council for Mutual Economic Assistance (COMECON) established by Moscow, tightening economic links among Soviet bloc countries.
April 4, 1949	North Atlantic Treaty signed, establishing North Atlantic Treaty Organization (NATO), the Western military alliance.
April 18, 1949	Republic of Ireland established; Northern Ireland region remains part of United Kingdom.
May 12, 1949	Berlin blockade lifted by Soviets; U.S. airlift continues until Sept. 30.
May 21, 1949	Federal Republic of Germany (West) established.
June 29, 1949	U.S. occupying forces withdrawn from South Korea.
Oct. 1, 1949	People's Republic of China proclaimed by victorious communist forces led by Mao Zedong after protracted civil war with Nationalists under Chiang Kai-shek.
Oct. 7, 1949	German Democratic Republic (East) established.
Dec. 14, 1949	Israel transfers capital to Jerusalem.

1950s

Feb. 9, 1950	In speech before Ohio County Women's Republican Club in Wheeling, W. Va., Sen. Joseph McCarthy sparks major "red scare" in U.S., claiming to have list of known Communists in State Department.
March 1, 1950	Chiang Kai-shek becomes president of Republic of China, based on Taiwan, where Nationalist Chinese establish government after losing mainland China to Communists.
March 8, 1950	Soviets announce successful development of atomic bomb.
May 8, 1950	Truman sends economic aid and establishes military mission in Vietnam, marking first step toward direct U.S. involvement.
June 25, 1950	North Korea invades South Korea, beginning Korean War.
June 27, 1950	UN Security Council approves military aid for South Korea; 15 nations will send troops as part of UN force.
June 30, 1950	U.S. troops under command of MacArthur enter Korean war.
Sept. 15, 1950	U.S. troops land at Inchon, South Korea; driving north, they reach the Yalu River on the Chinese border on Oct. 24, provoking Chinese counterattack which forces MacArthur to retreat into the south.

Sept. 30, 1950	Truman approves NSC 68, recommending an unprecedented, ambitious peacetime program to expand U.S. conventional and nuclear forces and to rearm American allies.
April 11, 1951	MacArthur, accused of insubordination, is relieved of command in Korea by Truman.
June 14, 1951	At U.S. Census Bureau, Univac I becomes first commercially manufactured computer operated in U.S.
Sept. 8, 1951	Peace treaty signed with Japan by 49 nations in San Francisco; U.S. and Japan agree to mutual security pact.
March 10, 1952	Cuban presidential candidate Fulgencio Batista seizes power and maintains authoritarian rule until 1959.
July 25, 1952	Adopting new constitution, Puerto Rico becomes self-governing commonwealth of U.S.
Oct. 2, 1952	Britain tests its first atomic device.
Oct. 31, 1952	First hydrogen bomb detonated by U.S. in the Marshall Islands.
Nov. 4, 1952	Republican Dwight D. Eisenhower defeats Adlai E. Stevenson in presidential election.
March 5, 1953	Stalin dies and is succeeded as Soviet premier by Georgi M. Malenkov.
June 17, 1953	Anticommunist riots in East Berlin lead to general strike; Soviet troops used to quell disturbances.
June 19, 1953	Ethel and Julius Rosenberg electrocuted for espionage, the first such peacetime execution in U.S. history.
July 27, 1953	After more than two years of negotiation, armistice ending Korean War signed at Panmunjom between UN forces and North Korea; border secured at 38th parallel with demilitarized buffer zone dividing North and South Korea; under terms of separate mutual aid pact, U.S. will provide South Korea with protective military presence and economic assistance for decades to come.
Aug. 12, 1953	Soviets explode hydrogen device.
Aug. 19, 1953	CIA sponsors coup overthrowing leftist Iranian Prime Minister Mohammed Mossadegh in response to Eisenhower administration concerns about rising Soviet influence in Persian Gulf; Mohammed Reza Shah Pahlevi reinstated on Peacock Throne, ruling until 1978 Islamic fundamentalist revolution.

Sept. 13, 1953	Nikita Khrushchev becomes first secretary of Soviet Communist Party; becomes premier in 1958, achieving preeminence in Soviet leadership.
Jan. 21, 1954	First nuclear-powered submarine, U.S.S. *Nautilus,* launched.
April 7, 1954	"Domino theory" first propounded by Eisenhower.
April 22, 1954	McCarthy accuses Secretary of the Army Robert Stevens of communist sympathies; Army counteraccusations lead to televised Army/McCarthy hearings.
May 7, 1954	At Dien Bien Phu, Vietnam, French forces suffer devastating defeat by communist nationalist forces under Ho Chi Minh.
May 17, 1954	Supreme Court decision in *Brown* v. *Board of Education* outlaws school segregation in U.S.
June 27, 1954	CIA-backed military revolt deposes leftist Guatemalan president Jacobo Arbenz.
June 27, 1954	First nuclear power station begins operation near Moscow.
July 20, 1954	After several years of fighting between French and Vietnamese anticolonial forces, Geneva Accords divide Vietnam at 17th parallel, establishing communist regime in north and noncommunist nationalist regime in south and providing for elections and reunification in two years; French withdraw from Indochina.
Oct. 22, 1954	Eisenhower authorizes training program for South Vietnamese army.
Oct. 23, 1954	Western alliance governments sign Paris Agreements, permitting West Germany to rearm and enter NATO.
Dec. 2, 1954	Senate votes to censure McCarthy.
Jan. 12, 1955	Secretary of State John Foster Dulles propounds nuclear strategy of massive retaliation.
April 17–24, 1955	At conference in Bandung, Indonesia, representatives of 29 less-developed nations launch nonaligned movement, dedicating themselves to conducting their affairs independently of superpowers.
April 29, 1955	Armed conflict breaks out between communist forces from North Vietnam and noncommunist forces from the south.
May 5, 1955	With end of U.S. occupation, West Germany becomes sovereign state.
May 14, 1955	Warsaw Pact alliance, Soviet bloc counterpart to NATO, formed ostensibly to offset West German remilitarization.

Oct. 26, 1955	South Vietnamese proclaim republic, rejecting elections stipulated in Geneva Accords.
Feb. 14, 1956	Khrushchev's denunciation of Stalin at opening of 20th Soviet Communist Party Congress marks beginning of period of political liberalization in U.S.S.R. and Eastern Europe.
April 23, 1956	Supreme Court outlaws racial segregation on all public transportation in U.S.
June 28, 1956	More than 100 people killed in anticommunist demonstrations in Poznan, Poland.
Oct. 29, 1956	Israel invades Sinai following Egypt's blockade of the Gulf of Aqaba; Britain and France intervene militarily on Israel's side on Oct. 31 after Egypt rejects cease-fire. On Nov. 5, British paratroopers recover Suez Canal, seized by Egyptian leader Gamal Abdel Nasser in July; Anglo-French action condemned by UN and U.S.; cease-fire takes effect Nov. 7, and UN peacekeeping force arrives Nov. 15.
Nov. 4, 1956	Soviet troops crush Hungarian uprising led by Imre Nagy, who had sought to create a Hungarian national communism distinct from Soviet model.
Nov. 6, 1956	Eisenhower reelected, defeating Stevenson again.
March 25, 1957	European Economic Community (Common Market) established.
Sept. 24, 1957	Eisenhower dispatches federal troops to Little Rock, Ark., to enforce desegregation of Central High School.
Oct. 4, 1957	*Sputnik I*, first man-made satellite, launched into earth orbit by U.S.S.R.
July 15, 1958	U.S. Marines land in Lebanon to restore order amid sectarian strife.
Jan. 1, 1959	Cuban dictator Batista flees as rebel forces under Fidel Castro sweep into nation's cities, establishing the first communist regime in Western Hemisphere.

1960s

Feb. 1, 1960	Black students denied service at a Greensboro, N.C., lunch counter begin sit-in protests.
Feb. 13, 1960	France becomes world's fourth nuclear power.
March 21, 1960	South African police open fire on unarmed black demonstrators in township of Sharpeville, killing 69 and wounding more than 200.

April 27, 1960	South Korean president Syngman Rhee resigns following massive anticorruption demonstrations in Seoul and U.S. rebuke for political repression; Park Chung Hee eventually succeeds to presidency.
May 1, 1960	American U-2 spy plane shot down over U.S.S.R.
Aug. 12, 1960	U.S. launches world's first communications satellite.
Nov. 8, 1960	John F. Kennedy defeats Richard Nixon in presidential election.
Jan. 3, 1961	U.S. severs diplomatic relations with Cuba.
March 1, 1961	Peace Corps founded.
March 13, 1961	Kennedy proposes Alliance for Progress, ambitious program for economic and social development in Latin America.
April 17–20, 1961	CIA-sponsored invasion of Cuba by anti-Castro exiles, staged at the Bay of Pigs, collapses in disarray.
May 4, 1961	"Freedom Riders," civil rights protesters, begin peaceful demonstrations in Birmingham, Ala.
May 31, 1961	South Africa breaks all ties with British commonwealth, becoming fully independent republic.
Aug. 15–17, 1961	Berlin Wall constructed, permanently closing off eastern sector of Berlin from traffic and visitors from the west.
Nov. 14, 1961	Kennedy sends first U.S. helicopter companies to Vietnam.
June 25, 1962	Supreme Court declares mandatory prayer in public schools unconstitutional.
Sept. 30, 1962	U.S. marshals escort James Meredith to University of Mississippi, which previously excluded him on racial grounds.
Oct. 22, 1962	Kennedy announces U.S. naval blockade of Cuba after installation of Soviet missiles is discovered; on Oct. 28, at brink of war, Khrushchev agrees to remove missiles; U.S. blockade ends Nov. 20.
Aug. 5, 1963	Limited Nuclear Test Ban Treaty signed in Moscow by foreign ministers of Britain, U.S.S.R., and U.S.
Aug. 30, 1963	"Hot line," direct crisis communications link between Washington and Moscow, goes into service.
Nov. 22, 1963	Kennedy assassinated in Dallas; Vice-President Lyndon B. Johnson becomes president.

May 28, 1964	Palestine Liberation Organization (PLO) founded.
July 2, 1964	Civil Rights Act of 1964 signed into law by Johnson; prohibits discrimination on basis of race, religion, or national origin in all forms of public accommodation.
July 18, 1964	Race riots erupt in Harlem and spread to other northern cities.
Aug. 7, 1964	Responding to allegedly unprovoked North Vietnamese attack on U.S. vessels on Aug. 2, Congress passes Gulf of Tonkin Resolution granting Johnson broad powers to defend U.S. forces; resolution later used to justify vast buildup of U.S. combat forces in Vietnam.
Oct. 15, 1964	Khruschev ousted as Soviet leader; Leonid Brezhnev assumes party leadership, and Alexei Kosygin becomes premier.
Oct. 16, 1964	China announces successful nuclear test, becoming world's fifth nuclear power.
Nov. 3, 1964	Johnson elected to full term as president, defeating conservative Republican candidate Barry Goldwater.
Dec. 3, 1964	Free Speech Movement at University of California in Berkeley holds sit-in demonstration in which 814 people are arrested, the largest mass arrest in U.S. to date; Berkeley protests spark campus unrest across the country.
Feb. 1, 1965	Rev. Martin Luther King, Jr., and 770 other civil rights demonstrators arrested in Selma, Ala., while protesting racial discrimination in voter registration.
Feb. 7, 1965	Johnson initiates bombing of selected targets in North Vietnam.
March 7, 1965	While marching to state capital in Montgomery, Ala., black protesters led by King are attacked in Selma by more than 200 state police; with protection of 3000 U.S. national guardsmen, marchers successfully reach Montgomery on March 25.
March 8, 1965	First U.S. combat troops land in Vietnam.
April 28, 1965	U.S. marines intervene in civil unrest in Dominican Republic.
Aug. 10, 1965	U.S. Voting Rights Act goes into effect; federal officials begin registering black voters in Alabama, Louisiana, and Mississippi.
Oct. 15, 1965	Protest march from Berkeley to Oakland army base in California, nation's first major antiwar demonstration.

Nov. 25, 1965	Bloodless coup establishes Gen. Joseph Mobutu as president of the Congo after five years of civil war; in expression of cultural nationalism, Mobutu later changes country's name to Zaire and his own name to Mobutu Sese Seko.
Dec. 17, 1965	Ferdinand Marcos elected president of Philippines; by early 1970s, after declaring martial law, he will exert dictatorial control over national life.
Feb. 3, 1966	Soviet craft achieves first soft landing on moon.
March 12, 1966	Indonesian nationalist leader and president, Sukarno, overthrown by Lieutenant-General Suharto, who establishes military dictatorship.
July 1, 1966	Citing American threat to country's independence of action and its need to develop own nuclear deterrent force, Charles de Gaulle withdraws France from NATO.
Jan. 27, 1967	Treaty prohibiting use of space for military purposes signed by 62 nations in ceremonies in London, Moscow, and Washington.
June 5, 1967	"Six-Day War" breaks out in Middle East; Israel captures and occupies Sinai, West Bank, and Golan Heights territories.
Aug. 8, 1967	Indonesia, Malaysia, Thailand, Philippines, and Singapore launch Association of Southeast Asian Nations (ASEAN) to promote regional economic and political cooperation.
Dec. 2, 1967	First human heart transplant performed in Cape Town, South Africa, by Dr. Christiaan Barnard.
Jan. 23, 1968	U.S.S. *Pueblo,* naval intelligence ship, seized off coast of North Korea by that country's armed forces; 83-man crew held until Dec. 22, 1968.
Jan. 30, 1968	Tet offensive by North Vietnamese and Viet Cong forces begins; although unsuccessful militarily, Tet has significant impact in turning Americans against Vietnam war.
March 16, 1968	U.S. army unit under Lt. William Calley massacres approximately 200 Vietnamese civilians in village of My Lai; revelation of massacre in fall of 1969 catalyzes public opposition to American involvement in war; Calley, who claims to have been following orders, given life sentence for murders in March 1971; sentence later reduced to ten years.
April 4, 1968	King assassinated in Memphis.
April 23, 1968	Columbia University shut down by massive student demonstrations and sit-ins.

May 10, 1968 — First peace talks between U.S. and North Vietnam begin in Paris.

May 22, 1968 — Major wave of worker-student unrest begins in France, nearly toppling government of Charles de Gaulle.

June 5, 1968 — Sen. Robert F. Kennedy assassinated in Los Angeles while campaigning in California Democratic presidential primary.

Aug. 20, 1968 — Soviet forces invade Czechoslovakia, suppressing reformist regime of Alexander Dubcek.

Nov. 5, 1968 — Nixon defeats Hubert H. Humphrey in presidential election.

Nov. 12, 1968 — Brezhnev enunciates the "Brezhnev doctrine," asserting Soviets' right forcefully to intervene in affairs of sovereign nations when an established communist regime is in danger of collapse or subversion.

March 2, 1969 — Soviet and Chinese troops clash for first time on mutual border; hostilities mark beginning of significant Sino-Soviet tension.

March 18, 1969 — U.S. commences secret bombing of Cambodia.

April 21, 1969 — British troops stationed in Northern Ireland called in to enforce order in wake of violent rioting between Catholics and Protestants.

June 8, 1969 — After conferring with South Vietnamese President Nguyen Van Thieu, Nixon announces beginning of phased withdrawal of U.S. troops from Vietnam; "Vietnamization" of war underway.

July 21, 1969 — Neil Armstrong is first man to walk on moon.

Sept. 1, 1969 — Capt. Muammar al-Qaddafi stages military coup in Libya and proclaims socialist Libyan Arab Republic.

Nov. 3, 1969 — "Nixon Doctrine," calling on U.S. allies to bear increased military burdens for their own defense and protection of American foreign interests, publicly enunciated for first time in televised address; doctrine had initially emerged in press briefings during Nixon's trip to the Pacific the previous spring.

Nov. 24, 1969 — U.S. and U.S.S.R. sign UN-sponsored Treaty on the Non-Proliferation of Nuclear Weapons.

1970s

Jan. 12, 1970 — Leaders of Biafran independence movement surrender to Nigerian government, ending catastrophic 31-month civil war.

May 4, 1970	Four Kent State University students killed by Ohio national guardsmen during campus protest against American armed incursion into Cambodia.
Sept. 28, 1970	Nasser dies of heart attack and is succeeded by Egyptian Vice-President Anwar Sadat.
Nov. 3, 1970	Salvador Allende, first Marxist to be democratically elected head of government in the West, sworn in as president of Chile.
Jan. 25, 1971	Ugandan president Milton Obote deposed in bloody military coup led by Maj. Gen. Idi Amin.
June 13, 1971	*New York Times* begins publication of "Pentagon Papers," Defense Department's classified internal account of Vietnam decision-making process. Nixon's Justice Department obtains federal court order interrupting publication on June 15; other newspapers begin to publish documents and are also hit by court rulings; Supreme Court rules against government on June 30, and publication resumes; Daniel Ellsberg, source of the documents, charged with espionage and theft of government property, but case is later dismissed.
Aug. 15, 1971	As part of package of emergency reform measures, Nixon announces end of convertibility of U.S. currency into gold in effort to free dollar for devaluation relative to other currencies, to slow speculation against dollar, and to close balance of foreign payments deficit.
Oct. 25, 1971	UN General Assembly votes to seat Communist People's Republic of China, replacing Taiwan.
Dec. 16, 1971	With Indian assistance, nationalist rebels in East Pakistan succeed in breaking away from Pakistan, establishing independent nation of Bangladesh.
Feb. 20, 1972	Nixon arrives in Peking, becoming first American president to visit People's Republic of China.
May 22, 1972	Nixon is first American president to visit Moscow.
June 17, 1972	Five employees of Nixon reelection committee arrested while breaking into Democratic National Committee headquarters in Watergate office complex in Washington.
Sept. 5, 1972	Members of Black September Arab terrorist group seize Israeli athletes at Munich Olympics; 11 hostages killed during day-long stand-off and subsequent gun-fight at Munich airport.
Oct. 3, 1972	SALT I and Anti-Ballistic Missile (ABM) arms limitation treaties signed by U.S. and U.S.S.R. in Washington.

Nov. 7, 1972	Nixon wins presidential reelection, defeating liberal Democrat George McGovern.
Jan. 27, 1973	North Vietnam, South Vietnam, Viet Cong, and U.S. sign Agreement on Ending the War and Restoring Peace in Vietnam, the Paris Peace Agreement.
March 29, 1973	Last U.S. troops withdrawn from Vietnam.
July 23, 1973	Asserting constitutional right to executive privilege, Nixon defies federal court order to hand over tape recordings of White House conversations to Justice Dpartment officials investigating Watergate scandal.
Aug. 15, 1973	U.S. bombing of North Vietnamese logistical targets in Cambodia terminated against president's wishes after Congress halts funding in effort to bar further U.S. military action in Indochina.
Sept. 11, 1973	Chilean military overthrows Allende.
Oct. 6, 1973	Egypt and Syria attack Israel, beginning Yom Kippur War.
Oct. 10, 1973	Vice-President Spiro T. Agnew resigns under pressure, pleading no contest to charges of tax evasion; Nixon selects House minority leader, Gerald R. Ford, to replace him.
Oct. 19, 1973	First oil embargo by member nations of the Organization of Petroleum Exporting Countries (OPEC) leads to debilitating oil shortages and economic disruption in nations dependent on imported oil.
Oct. 20, 1973	In "Saturday Night Massacre," Attorney General Elliot Richardson and Deputy Attorney General William Ruckelshaus resign and Special Prosecutor Archibald Cox is fired following disagreements with Nixon over Justice Department's role in Watergate investigation.
April 25, 1974	Portuguese premier Marcello Caetano ousted; revolutionary turmoil continues for two years, contributing to dissolution of Portugal's African colonial empire. Government signs agreement freeing Portuguese Guinea on Aug. 26.
May 9, 1974	U.S. House Judiciary Committee votes to impeach Nixon.
May 18, 1974	India explodes nuclear device, becoming world's sixth nuclear power.
July 15, 1974	Coup d'etat by Greek officers in Cypriot National Guard ousts government of Archbishop Makarios in Cyprus, prompting July 20 invasion by Turkey to protect Turkish Cypriot minority; fighting continues through end of summer, when U.N.-sponsored cease-fire holds.

Aug. 9, 1974	Nixon resigns under threat of impeachment; Ford sworn in as first nonelected president in U.S. history.
Sept. 4, 1974	U.S. establishes diplomatic relations with East Germany.
Sept. 8, 1974	Ford grants Nixon full pardon for any wrongdoing in Watergate affair.
Sept. 12, 1974	Radical military leaders depose Ethiopian emperor Haile Selassie after 58-year reign; socialist state declared on Dec. 20.
April 16, 1975	U.S.-backed Cambodian government of Lon Nol falls to communist Khmer Rouge, ending five-year civil war; under ultra-radical new regime led by Pol Pot, 1 to 2 million people will die by execution, starvation, and disease over next four years.
April 30, 1975	Saigon captured by North Vietnamese after surprise offensive overwhelms South Vietnamese army; South Vietnamese president Duong Van Minh surrenders unconditionally as Americans remaining in Saigon are evacuated by helicopter.
May 12, 1975	Cambodian forces seize U.S. cargo ship *Mayaguez* along with 40-man crew; rescue mission on May 14 frees crew, but gun-fight and helicopter crash result in deaths of 41 U.S. marines.
Aug. 1, 1975	At conclusion of Conference on Security and Cooperation in Europe, 33 European nations, Canada, and U.S. sign nonbinding pact freezing postwar borders, broadening detente, and pledging respect for human rights.
Aug. 23, 1975	After 20 years of war and two of coalition government, Vietnamese-backed communist Pathet Lao establish exclusive control over Laos.
Nov. 10, 1975	Portugal's African empire fully dissolved as Angola gains independence after 15 years of civil war; within three months, Soviet-backed Popular Movement for Liberation of Angola emerges victorious from power struggle among guerrilla factions.
Dec. 15, 1975	UN General Assembly passes resolution proclaiming 1976–85 period the "U.N. Decade for Women: Equality, Development, and Peace"; economic, social, and political objectives of U.N. Decade previously set forth in "World Plan of Action" approved at July 1975 Mexico City conference on women.
May 28, 1976	Treaty prohibiting underground nuclear explosions signed by Ford and Brezhnev.

June 16, 1976	Protest by more than 10,000 black students in South African township of Soweto begins.
July 2, 1976	Vietnam formally reunited as one nation, under communist rule.
Sept. 9, 1976	Mao Zedong dies, sparking succession struggle eventually won by Hua Kuofeng.
Nov. 2, 1976	Democratic candidate Jimmy Carter defeats Ford in presidential election.
Jan. 21, 1977	Carter grants full pardon to Vietnam War era draft resisters.
Sept. 7, 1977	Panama Canal treaties signed, establishing timetable for turning over eventual control of canal to Panama and guaranteeing its perpetual neutrality.
Nov. 9, 1977	Egyptian president Sadat travels to Jerusalem to meet with Israeli prime minister Menachem Begin; two weeks later Sadat breaks diplomatic relations with several hard-line Arab nations that denounced meeting with Begin.
April 27, 1978	Afghan president Mohammed Daud Khan ousted by military; Communist party leader Noor Mohammed Taraki assumes power.
Sept. 17, 1978	With Carter mediating, Sadat and Begin reach Egyptian-Israeli peace agreement at American presidential retreat in Camp David, Md. Camp David Accords signed on March 26, 1979.
Oct. 16, 1978	Polish Cardinal Karol Wojtyla becomes Pope John Paul II.
Jan. 1, 1979	U.S. and China establish diplomatic relations.
Jan. 7, 1979	Vietnamese invasion forces bring down Cambodian government of Pol Pot and occupy country, putting in place pro-Hanoi Cambodian communist regime under Heng Samrin.
Jan. 16, 1979	Shah and family leave Iran as regime collapses under pressure of Islamic fundamentalist revolution; Ayatollah Ruhollah Khomeini, fundamentalist spiritual leader, returns to Iran at the end of the month, taking control of government and imposing strict Islamic code upon nation's social and political life.
Feb. 17, 1979	Approximately 250,000 Chinese troops invade Vietnam in response to border clashes and aggression in Cambodia; both sides suffer massive casualties before Chinese troops withdraw on March 15.

March 28, 1979	First major U.S. nuclear power plant accident occurs at Three Mile Island reactor near Harrisburg, Pa.
April 11, 1979	Force of Ugandan exiles and Tanzanian troops occupies Ugandan capital of Kampala, ending brutal eight-year regime of Amin, who flees to Libya.
May 3, 1979	In landslide victory for Conservative party, Margaret Thatcher becomes Britain's first woman prime minister.
June 18, 1979	Carter and Brezhnev sign SALT II Treaty in Vienna.
July 17, 1979	Insurrectionary forces led by Sandinista National Liberation Front (FSLN) take over Nicaraguan capital of Managua, ending 50-year dictatorial rule of Somoza family; President Anastasio Somoza flees.
Oct. 1, 1979	U.S. Canal Zone formally dissolved as Panama takes control of territory in accord with 1977 treaties; canal itself scheduled to be turned over to Panama on Dec. 31, 1999.
Nov. 4, 1979	U.S. embassy in Tehran stormed by Iranian militants who seize 66 American hostages; 52 ultimately held for almost 15 months; among other demands, militants call for Shah to be returned to Iran to stand trial; after leaving U.S., Shah dies in Egypt in July 1980.
Nov. 21, 1979	Islamic fundamentalists burn U.S. embassy in Pakistan.
Dec. 12, 1979	Under so-called two-track decision, NATO governments, excluding France and Greece, agree to deploy 572 new intermediate-range nuclear missiles in Europe by end of 1983 while simultaneously pursuing agreement with U.S.S.R. on controlling all intermediate-range weapons based in Eastern and Western Europe; projected deplayment of new missiles, first such weapons capable of reaching targets inside U.S.S.R. from European soil, prompts massive antinuclear demonstrations in Europe during early 1980s.
Dec. 25, 1979	Following Moscow-engineered coup by Babrak Karmal, Soviet forces invade Afghanistan and shore up new Marxist government; Carter administration recalls U.S. ambassador to Moscow, imposes embargo on grain and technology exports to Soviets, and asks Senate to delay ratification of SALT II treaty; U.S. Olympic Committee, under pressure from White House, later votes to boycott 1980 Olympic Games in Moscow.

1980s

April 6, 1980	After thousands of Cubans storm Peruvian embassy seeking asylum, Castro opens port of Mariel in Cuba, allowing

thousands of small boats to leave for U.S.; before port is closed at the end of September, approximately 125,000 Cubans, many convicted criminals and mental patients, will have emigrated.

April 17, 1980	Ending 90 years of white rule and 8 years of interracial civil war, independent, majority-ruled nation of Zimbabwe created from white-ruled Rhodesia; nationalist leader Robert Mugabe, whose party had won landslide parliamentary electoral victory in March, sworn in as prime minister.
April 24, 1980	Eight U.S. commandos die in aborted attempt to rescue American hostages in Iran.
May 21–27, 1980	An estimated 1000 people killed by South Korean military forces during uprisings in Kwangju spurred by detention of opposition leader Kim Dae Jung.
July 19, 1980	Summer Olympic Games open in Moscow without participation of U.S. and other boycotting Western nations.
Aug. 30, 1980	Eastern Europe's first independent postwar trade union, Solidarity, formed in Poland.
Nov. 4, 1980	Republican Ronald Reagan defeats Carter in presidential election.
Jan. 20, 1981	Remaining 52 U.S. hostages released in Iran moments after Reagan is sworn in as president.
March 30, 1981	Reagan shot in Washington by John W. Hinckley, Jr., but recovers quickly.
May 10, 1981	Francois Mitterrand becomes first Socialist Party candidate elected president of France.
May 13, 1981	Pope John Paul II shot in Rome by Mehmet Ali Agca, a Turk; Italian investigators later charge three Bulgarians with plotting the attempt.
Oct. 6, 1981	Sadat assassinated in Cairo by members of radical Muslim sect; Vice-President Hosni Mubarak succeeds to presidency.
Oct. 22, 1981	Leaders of 22 nations meet in Cancun, Mexico, to discuss ambitious agenda of issues concerning economic cooperation between industrial countries of the north and developing countries of the south.
Dec. 13, 1981	Declaration of martial law in Poland results in banning of Solidarity trade union.
Jan. 8, 1982	Justice Department and AT&T reach settlement in which latter agrees to divest itself of 22 regional Bell System companies, breaking its monopoly over nation's phone services.

April 2, 1982	Suspending negotiations, Argentina invades contested Falkland (Malvinas) Islands in South Atlantic, provoking British prime minister Thatcher to dispatch naval task force which retakes islands on June 14; Argentine defeat hastens fall of military junta, leading eventually to return of civilian democratic rule.
June 6, 1982	Israel invades Lebanon in effort to secure its northern border region from terrorist and artillery attacks and to crush PLO's military apparatus.
June 30, 1982	Legal deadline passes for ratification of Equal Rights Amendment to U.S. Constitution; only 35 of required 38 states have ratified ERA, which would prohibit discrimination on basis of sex.
Aug. 25, 1982	U.S. marines land in Lebanon as part of multinational peacekeeping force; leaving after two weeks, they return for extended stay on September 29, along with French and Italian contingents, after massacre of more than 800 Palestinian men, women, and children by Lebanese Christian militiamen at Sabra and Shatila refugee camps in Beirut.
Sept. 12–13, 1982	Major shake-up of Chinese leadership results in election of Deng Xiaoping as chairman of newly formed Central Advisory Committee to the Communist Party, consolidating his power in both party and government.
Nov. 10, 1982	Brezhnev dies; Yuri Andropov, former KGB chief, succeeds him as general secretary of Soviet Communist Party on Nov. 12.
March 23, 1983	In nationally televised speech, Reagan outlines plans for funding research on Strategic Defense Initiative, a space-based missile defense system familiarly known as "Star Wars."
April 18, 1983	U.S. embassy in Beirut bombed, killing more than 50.
May 20, 1983	Car bomb detonated outside South African air force headquarters in Pretoria kills 18 in most serious single terrorist attack to date by black resistance forces.
July 21, 1983	Martial law repealed in Poland.
Sept. 1, 1983	Korean airliner with more than 300 passengers aboard shot down over Sea of Japan by Soviet fighter plane after entering Soviet airspace.
Oct. 3, 1983	Barracks of U.S. marines participating in Lebanon peacekeeping force destroyed by car bomb, killing 241 servicemen; 58 die in similar attack on French headquarters on same day.

Oct. 25, 1983 — Citing threat to U.S. security interests and to lives of American medical students, U.S. forces invade Caribbean island of Grenada, deposing extreme left-wing faction that had ousted and killed Prime Minister Maurice Bishop one week before; hundreds of Cuban construction workers and soldiers are routed and captured during subsequent days; new elections held under U.S. sponsorship in 1984.

Oct. 30, 1983 — Radical Party leader Raul Alfonsin elected president of Argentina, defeating a candidate of the influential Peronist movement and ending eight years of repressive military rule.

Nov. 2, 1983 — White voters approve new South African constitution providing representation in separate chambers of parliament for Indians and mixed-race "Coloureds," but not for black majority.

Nov. 23, 1983 — Soviet delegation walks out of Intermediate Nuclear Force reduction talks in Geneva; one month later, Soviets refuse to set date for resumption of Strategic Arms Reduction Talks (START).

Dec. 30, 1983 — Despite substantial popular opposition and protest, first nine Pershing II missiles called for in NATO's 1979 nuclear modernization scheme become operational in West Germany.

Feb. 9, 1984 — Andropov dies and is succeeded by Konstantin Chernenko.

April 6, 1984 — Reports disclose CIA role in directing mining of Nicaraguan ports; Sandinista government appeals to World Court, but Reagan administration announces in advance that it will not recognize court's jurisdiction in this case.

May 6, 1984 — José Napoleon Duarte, moderate candidate favored by U.S. government, elected president of El Salvador, defeating right-wing candidate Roberto d'Aubuisson.

July 19, 1984 — Democratic National Convention in San Francisco nominates Geraldine Ferraro as first woman vice-presidential candidate of major party in U.S.

July 28, 1984 — Summer Olympics open in Los Angeles, with U.S.S.R. leading boycott of communist nations.

Sept. 20, 1984 — New U.S. embassy in West Beirut bombed.

Sept. 25, 1984 — Resolution banning visits by nuclear-armed warships introduced in New Zealand parliament, sparking controversy with U.S. and threatening future of 1951 tripartite Australia–New Zealand–U.S. security agreement commonly known as ANZUS.

Oct. 20, 1984	Chinese Communist Party issues major document approving Deng's program for liberalization and decentralization of Chinese economy.
Oct. 31, 1984	Following series of protracted, violent clashes between Sikh militants and Indian police and military in Punjab state, Indian Prime Minister Indira Gandhi assassinated by Sikh members of her own personal guard, her son Rajiv succeeds her in office.
Nov. 6, 1984	Defeating former Vice-President Walter F. Mondale, Reagan reelected president.
Nov. 20, 1984	Red Cross representatives from North and South Korea meet to discuss reunification of families separated by Korean war.
Dec. 9, 1984	With passing of deadline for signatures of UN Treaty on the Law of the Sea, U.S. fails to join 138 countries and 21 international organizations which have put their names to the document.
March 10, 1985	Chernenko dies; Mikhail Gorbachev, first member of successor generation of Soviet leadership, ascends to power.
March 12, 1985	Comprehensive U.S.-Soviet arms control negotiations involving linked talks on strategic, intermediate-range, and space-based weapons systems resume in Geneva after more than one-year hiatus.
June 14, 1985	TWA flight hijacked and American passengers taken hostage by Lebanese Shiite Muslims demanding release of Shiite prisoners held by Israel; negotiations lead to freeing of hostages 17 days later after personal intervention by Syrian president Hafez al-Assad.
July 2, 1985	Andrei Gromyko, a major architect of postwar Soviet foreign policy, named to ceremonial office of president of U.S.S.R. after 28 years as foreign minister; Eduard Shevardnadze becomes foreign minister; Reagan and Gorbachev agree to November 1985 summit meeting in Geneva, first such encounter for both leaders.
July 20, 1985	In face of escalating unrest in black townships, white government declares state of emergency in certain regions of South Africa; violence intensifies as police arrest more than 1500 opposition figures under new law-enforcement guidelines.

Index

DATE DUE
